T0246315

Driving Behavior

In the U.S., drivers over the age of 65 now account for nearly 20% of licensed drivers. This number will increase by 25% to nearly 70 million by the year 2030. Some of these older drivers may not be capable of operating their vehicles safely in all conditions. The book investigates the key aspects of driving tasks and their relation to the sensory, perceptual, motor, and cognitive processes. Effects on driving performance, including aging, are described with a view toward improving future vehicle and road design as well as driver training and evaluation.

This title:

- Presents a comprehensive, quantitative analysis of human physical and mental processes to driving behavior.
- Showcases recent review and analysis of driver, vehicle, and road environment design factors.
- Discusses the fundamentals of driving behavior in vehicle control and guidance, collision avoidance, and hazard perception.
- Examines the effects of design issues on vehicles and road environments.
- Highlights specific quantifiable attributes of physical and mental functions related to driving approaches.

Written for professionals in diverse fields including ergonomics, health and safety, human factors, transportation engineering, and automotive engineering, this book is the essential guide to driving practices and habits. Its appeal will extend to those involved in vehicle design, roadway environments, driver training, and regulatory agencies.

Alfred T. Lee has been involved in the conduct of research in human–systems integration for more than 40 years. His principal research focus has been developing an understanding of how human capabilities and limitations affect the way in which complex systems function. This understanding can be used to develop designs that maximize the influence of those attributes of the human operator that can enhance the efficacy of system operation while minimizing the influence of those aspects that do not. This is particularly important for system safety as designs that do not properly integrate the human operator inevitably reduce operational safety.

Driving Behavior

Managing Resources in a Complex Task

Alfred T. Lee

CRC Press
Taylor & Francis Group
Boca Raton London New York

CRC Press is an imprint of the
Taylor & Francis Group, an **informa** business

Designed cover image: © Shutterstock

First edition published 2024
by CRC Press
2385 NW Executive Center Drive, Suite 320, Boca Raton FL 33431

and by CRC Press
4 Park Square, Milton Park, Abingdon, Oxon, OX14 4RN

CRC Press is an imprint of Taylor & Francis Group, LLC

ISBN: 9781032431802 (hbk)
ISBN: 9781032593654 (pbk)
ISBN: 9781003454373 (ebk)

DOI: 10.1201/9781003454373

Typeset in Times New Roman
by Deanta Global Publishing Services, Chennai, India

Contents

Preface

There are few skills that, like driving, are shared almost universally among adult humans. Hundreds of millions of people learn to drive a motor vehicle of one type or another. In the U.S., there are approximately 230 million licensed drivers, with 3 million new drivers licensed each year. By 2030, about one in four of these drivers will be 65 years of age or older.

The ubiquity of licensed drivers, both in the U.S. and in other developed countries, means that driving skill is being acquired and exercised by individuals with a wide range of physical and mental resources. Yet they are able to perform the task of driving routinely in a relatively safe manner. Driver resources, including sensory, perceptual, perceptual-motor, and cognitive, are exercised through the development of skills which focus selected resources on a given driving task. An understanding of how these resources are utilized in driving in a wide variety of driving scenarios is needed in order to improve vehicle and road system design as well as driver training programs.

The chapters in this book are organized to provide an overview of basic driver resources, both physical and mental; a discussion of how these resources are deployed in essential driving tasks; and how the resource management framework can improve the driving task environment. In Chapter 1, a detailed description of driver resources is provided, including their operating limits and how they affect driving behavior. Chapter 2 addresses how driver resources are used in the essential tasks of vehicle speed and directional control. The critical task of collision avoidance is reviewed in Chapter 3, with particular emphasis on key pre-attentive and consciously controlled visual processes. In contrast to the imminent hazard of collisions, Chapter 4 discusses the driver's ability to detect, classify, and respond to potential road hazards, that is, hazards that may evolve in the near future. Also discussed is the value of hazard perception training and testing. Chapter 5 addresses the problem of multitasking and attention and how drivers deal with task demands that evolve from having more than one task to perform at a time. Given the growing proportion of older drivers in the population, Chapter 6 describes how normal, healthy aging impacts driver resources and how older drivers respond to these changes. Finally, Chapter 7 describes how the resource management approach to driving differs from other approaches and how resource management mechanisms such as the automatization for driving skills,

vehicle automation such as alerts and warnings, and road design elements such as enhanced surface texturing and marking can improve
driver resource allocation and efficacy in resource expenditure.

Alfred T. Lee, Ph.D.
Beta Research and Development
Los Gatos, CA

1 Sensation, Perception, and Cognition in Driving

A driver brings to the driving task a set of sensory systems, visual and non-visual, as well as perceptual and cognitive processing capabilities. The sensory systems are determinant about the kind of data the driver is capable of acquiring in support of the driving task. While the driving task is predominantly a visual one, the driver also receives information through non-visual channels. These include auditory sensations and sensations resulting from the physical motion of the vehicle, the sensory feedback from the positioning of limbs, and the tactile sensation from the driver's use of vehicle controls. These raw data are converted to information (percepts) useful in the performance of driving tasks. Distance percepts, for example, are derived from rudimentary visual cues to distance such as foreground texture-density. Some percepts can, in turn, elicit perception-action sequences. For example, the rapid expansion of the retinal size of any approach object or "looming" will typically elicit a collision avoidance response (e.g., emergency braking). Since these perception-action sequences are often purely reactive, they reduce the demand on higher-level, cognitive processes. Cognitive processes include selective attention, working, and long-term storage and retrieval, necessary for planning and decision-making. Situation awareness, the knowledge of the state of the driver's vehicle, and the surrounding environment rely on cognitive processes. The process of identification and classification of potential hazards (see Chapter 4) also relies on cognitive processes as do tasks such as way-finding and navigation. These cognitive processes are resource intense as they require focused attention, which is limited to one task at a time.

Each of the sensory, perceptual, and cognitive processes used in driving needs to be decomposed into their functional subsystems to determine how they serve as resources in support of the driving task.

VISUAL SYSTEM

The visual system of the driver is arguably the most important of the sensory systems. A variety of driver tasks are supported by the visual system including vehicle guidance and speed control, hazard detection and collision avoidance, road signage recognition and interpretation, and many

DOI: 10.1201/9781003454373-1

others. Several key subsystems support the visual system of a driver and define the limits and capabilities of driver vision.

MACULA AND FOVEA

The areas of the eye most relevant to the ability to resolve details in a visual scene are the macula and fovea. The macula covers some 18 deg of the center of the eye, plus or minus 9 deg from the fovea axis or center of the eye. This is where the visual acuity (VA) of the eye is the greatest, especially during daylight or photopic[1] operations. Photopic vision is available when the luminance of an object is above 3 cd/m². The eye operates optimally with regard to VA and color perception at luminance levels above 400 cd/m².

The highest concentration of photoreceptors is in the fovea or very center of the macula. This area covers about 5 deg of the center of the eye with the highest concentration of receptors covering 1 to 2 deg. It is the latter area of the eye which is tested by licensing in the use of a Snellen Chart or similar instruments. Thus it is the measure of the critical detail of an object such as the gap in the letter of a road sign. Snellen acuity of 20/40 (6/12 m) is a measure of the ability to detect a gap of about 2 arc min of the visual angle subtended by this critical detail.

From the center of the eye, the density of photoreceptors drops, along with VA, very rapidly. At 10 deg off the foveal axis, the loss of about 85% of photoreceptors means that much of the VA of the eye during photopic driving conditions is lost. However, areas beyond 10 deg have sufficient photoreceptors to sense gross object movements such as vehicles or pedestrians. As will be seen, these gross movement detections are important in the driver's estimate of self-motion and hazard detection.

MESOPIC VISION

Driving under night conditions means that the available luminance is less than the 3 cd/m² needed for photopic vision but is still more than the threshold of scotopic vision at 0.01 cd/m². As there is always some light available in driving, true scotopic (sometimes termed "night vision") does not occur. The range of luminance within mesopic vision will vary with road conditions, particularly between urban and rural environments. (Light luminance levels that are substantially beyond luminance at which the driver has become adapted or *glare* can result in significant impairment, especially during mesopic driving conditions.)

In mesopic vision, the photoreceptors used in photopic vision (cones) are no longer functioning at optimum levels of efficiency. Moreover, another class of photoreceptors (rods) now becomes active. These interact with and even inhibit cone receptors. This means that the high level of VA within the macula is no longer available to the driver.

The distribution of these photoreceptors is very different. Unlike the high density of cone receptors in the macula, rod receptors are predominantly outside of the macula and are completely absent in the fovea within 3 to 4 deg of the foveal axis. This means that the VA of critical object details measured by Snellen acuity under mesopic conditions is exclusively a measure of cone receptors operating suboptimally.

The effects on VA at such low luminance levels are significant. In studies of young observers with better than 20/20 vision, declines in VA under mesopic conditions reduced to 20/66 within 60 sec (Hiroaka et al., 2015). As will be seen, even small increases in luminance levels under mesopic conditions can be of great benefit to drivers.

COLOR VISION

Luminance levels not only impact the VA of drivers but also the ability to perceive colors accurately. As luminance levels decline with darkness, a driver's spectral sensitivity shifts to a shorter wavelength. The longer-wavelength (red-orange) sensitivity shifts to the shorter-wavelength (indigo-violet) area of the color spectrum. This Purkinje shift, as it is called, is one reason that color perception tends to be inaccurate in night driving. The trichromaticity (red, green, and blue) of photopic vision provided by cone receptors is slowly overtaken by the achromatic rod receptors as luminance levels decline to mesopic vision.

As traffic signaling and road signage systems have historically used redundant features, the impact of color vision inaccuracies is substantially reduced. For example, traffic signal lights are generally ordered top to bottom: red, amber, and green. The position, not just the color, of the active light tells the driver whether a stop, caution, or proceed action is required. Similarly stop signs, as well as being red in color, are lettered, octagonal in shape, and are always at an intersection. These redundancies in design should always be followed to avoid color confusion at low luminance levels.

Other than color vision decline with mesopic vision, genetic anomalies especially in males (about 8 %) result in poor red–green discrimination. Red–green discrimination due to its use in traffic signals is particularly problematic if those signals are not accompanied by redundancies in the design. This might occur in protected turns either crossing traffic in a left turn or merging traffic in a right turn where the signal light is in isolation from other signals. A variety of genetic anomalies and diseases affect color discrimination other than gender and should be considered in the design process. Most notably, the aging process also impacts color discrimination (see Chapter 6). Age may be the most important demographic consideration now and in the future, given the growing proportion of older drivers.

CONTRAST SENSITIVITY

It should be noted that VA is not the only important measure of driver vision. The luminance contrast between an object or an object's features and its immediate background affects the ability of a driver to identify that object or its features. The formula for luminance contrast or Weber Contrast is:

$$L - L_b / L_b$$

where L is the luminance of the object or feature and L_b is the luminance of the background. In a road sign, for example, luminance contrast is measured by comparing the luminance of the edge boundary of the sign with the sign's immediate background. The background luminance can vary widely between rural and urban environments and with the light available. For the road sign details, the lettering and other components are contrasted against the sign's background. The *legibility distance* of road signs is the distance at which the road sign can be read by the driver. It is determined, in large part, by the degree of contrast between the luminance of these details and the sign's background luminance and the contrast sensitivity of the driver.

Another form of contrast is based on color. *Color contrast sensitivity* is becoming more important as road signage moves away from simple black or white lettering on a white or colored background to combinations of colored letters or other details on a colored background and colored sign edge boundaries. In physical terms, color contrast is maximized when the object and background difference in wavelength or hue is maximum. Thus, red lettering on a purple background will have maximum color contrast. However, color is perceived by the driver along dimensions not only of wavelength or hue but of brightness and saturation as well. Thus, maximizing differences in color saturation and brightness as well as hue will maximize perceived color contrast.

In practical terms, luminance and color contrast should be considered together when addressing the design of road signage. In mesopic driving conditions, the problem of luminance and color contrast is amplified largely due to the Purkinje shift and the effect of low luminance on driver VA. Road signage legibility distance in mesopic conditions is determined largely by the ability of vehicle headlights to illuminate the sign. The differential luminance of the sign contents such as lettering and other details likely plays a larger role than it would under photopic conditions due to the poorer functioning of the driver's color-sensitive, cone photoreceptors under these driving conditions. Thus, dependence on purely color contrast for object discrimination may be successful under photopic conditions but may lead to reduced legibility distances under mesopic conditions.

Contrast sensitivity of a driver is particularly important in low luminance conditions and when image details are small. Thus, low luminance photopic conditions (e.g., 15 cd/m²) and all mesopic conditions rely on driver VA and contrast sensitivity. As a matter of course, both should be measured periodically especially if the driver is experiencing problems driving during low luminance photopic or under mesopic conditions. A contrast sensitivity function (CSF), which includes luminance values across a range of object visual angles, is a desirable improvement over separate VA and contrast sensitivity measures as it provides both in one function.

STEREOACUITY

By virtue of the fact that humans have two eyes situated beside one another, the resultant binocular vision results in the fact that each eye has a slightly different image of the same object, i.e., binocular disparity. The resultant difference in these two images results in *stereopsis* or the sense of perceived depth or distance of the object from the observer. Stereoacuity is a measure of the degree to which the observer can detect binocular disparity. Normal stereoacuity in an adult is about 20 arc sec. This ability to detect such small differences is important in judging the distances of objects close to the observer. For the driver, stereopsis is effective within 2 m and is most effective within 1 m of an object. Thus, the effectiveness of stereopsis is largely within rather than outside of the vehicle. While some stereopsis is evident at longer distances, the availability of stronger, monocular visual cues to distance tends to minimize its effectiveness in distance perception (see Distance Perception section below).

VISUAL FIELD OF VIEW

Thus far the description of the visual system of the driver has largely been for subsystems within the central area of the eye. The functional area of the eyes extends much further than the macula and fovea. The full visual field of view (FOV) extends about 180 deg horizontally for both eyes combined with about 40 deg of overlap between the two eyes. The vertical FOV extends about 50 deg above to 75 deg below the line of sight in the vertical. Much of the FOV is blocked vertically by the frame of a vehicle, however. For passenger cars only 30 to 40 deg of vertical FOV is available. In these vehicles, about 6 to 7 deg of the horizontal FOV of the driver is blocked by the left or A pillar and 2 to 3 deg of the FOV for the right or B pillar which supports the car's roof. These values vary somewhat depending on the particular vehicle make and model.

As noted earlier, the distribution of photoreceptors in the eye varies depending on whether photopic or mesopic driving conditions are involved. Under photopic conditions, the visual periphery is only very sparsely populated by cone photoreceptors. It is important, therefore, to understand the relationship between the driver's functional FOV and the roadway. The two areas of interest are the macula and fovea high-resolution areas of vision, which extend, respectively, 5 and 2 deg out from the vehicle for each eye. The combined high resolution for both eyes is thus 10 deg for the macula and 4 deg for the fovea.

As a practical example of the differences in the performance of the different areas of photoreceptor coverage, the likelihood of detection of a small object entering the roadway under photopic conditions will be calculated. A common tennis ball is 2.55 deg in diameter and is colored an optic yellow. It would therefore have both high luminance and high color contrast against the roadway surface. It is also a known precursor to a hazard event as it is likely to be followed by a human or animal in rapid pursuit. A driver with 20/40 VA would be able to detect the ball entering the roadway at a distance of about 369 ft (112 m), but only if it were detected with the very highest density of the foveal portion of the eye, i.e., within ±1 deg off the foveal axis. Even a slightly higher off-axis view of 5 deg would reduce the detection distance to 147 ft (45 m). An off-axis view of 9 deg further reduces the detection distance to 92 ft (28 m). Translated into the time available for the driver to stop the vehicle at 30 mph, only a slight off-axis detection of the object reduces the time to stop from about 8 to only 2 sec.

Fortunately, the eyes are capable of moving in their respective sockets. The maximum horizontal movement is about 45 deg though a more typical movement would be closer to 30 deg left or right. The highest performance of the eyes in detecting objects on the road is achieved only by visually scanning and searching. Driver visual scanning behavior will be discussed in more detail in Chapter 4.

VISUAL MOTION PERCEPTION

Beyond the macula, the eye is not capable of discerning any detail in photopic conditions. It is, however, capable of sensing motion. The motion can either be a moving object when the vehicle is stopped or the movement of the entire visual field when the vehicle is in motion. In the former case, the object can serve as an alerting stimulus for a potential hazard. In the latter case, the movement of elements in the whole of the visual field or *optic flow* gives the driver a strong sense of self-motion. (An illusory self-motion or *vection* can occur even when the vehicle is stopped. This might occur when the driver is parked next to a vehicle that backs out. Computer-generated visual field motion is also the means by which self-motion is usually provided in fixed-base driving simulators.)

The visual periphery plays an important role in understanding how drivers estimate vehicle speed when not attending to the speedometer. The optic flow field that envelopes the vehicle is interpreted by the driver's visual system at a primitive, pre-attentive level. At this level, the driver is not aware of the impact of the flow field on their perception of speed. The variation of perceptible elements in the flow field affects the degree of self-movement. The greater the density of the elements, the greater will be the sense of self-motion. The driving environment and lighting conditions will then determine the degree to which the visual flow field affects the driver's perception of self-motion.

Mesopic vision involves both types of photoreceptors. It was noted above that VA declines under mesopic conditions. However, the distribution of rod receptors outside of the macula is much higher than that of cone receptors. This would suggest that motion perception under mesopic conditions would be better than under photopic conditions. However, this is not the case. Mesopic motion perception in fact declines when compared to photopic conditions (Bilino et al., 2008; Yoshimoto et al., 2016). Despite the high density of rods in the visual periphery, their interaction with cone receptors results in a decline, not an increase, in motion perception sensitivity. As will be seen in the next chapter, low luminance in driving conditions has a strong effect on how well drivers can estimate their vehicle's speed.

EDGE RATE

Movement of the visual field is not the only means by which a driver can estimate vehicle speed. Self-motion can also be derived from the central visual field through the visual cue of *edge rate*. Edge rate refers to the rate at which edges of objects pass across a fixed point on the vehicle body. Typically, this is the windshield though other fixed parts could serve the same function. Edge rate is known to provide a strong sense of self-motion, even exceeding that of global field motion (Larish and Flach, 1990). An example of edge rate in driving is the rate at which the fore and aft edges of discontinuous lane markings cross the base of a driver's windshield. The actual edge rate is measured by the number of edges that cross the windshield each second. Edge rate will vary not only with vehicle speed but also with the length of and separation between markings. Further discussion of the edge rate cue to self-motion will be provided in Chapter 2.

DISTANCE PERCEPTION

One of the more critical capabilities of a driver is their ability to judge distance. Distance perception plays a role in a variety of maneuvers including the left turn in front of traffic, passing with oncoming traffic, and judging the safe distance to follow other traffic.

The perception of distance has been studied for many decades. Only a summary of what is known about distance perception will be provided here. Most of these studies are of static distance judgments, that is, when both the observer and the observed object are in a fixed position. Distance perception is, of course, important in judging the distance of moving objects, particularly those that are moving toward the driver.

Visual Cues to Distance

Analyses of the use of visual cues in distance judgments have revealed a number of elements within the optic array that are used by observers to judge the distance to objects and the relative distance of objects to one another (Cutting and Vishton, 1995). The relative size of an object is one such cue. Provided the object is a familiar one, an observer can use the perceived (or optical) size as a cue to distance. The smaller the optical size, the further the object will be perceived to be from the observer. Similarly, if two familiar objects in the same class (e.g., a passenger car) are viewed at a distance, the object with the relatively larger optical size will be seen as closer. This *size–distance invariance* is the most commonly used visual cues to distance.

Objects which are seen to visually overlap a second object in the visual field will be perceived as closer to the observer. This is true if only a portion of the objects overlap. Even if the relative optical size is greater as might be the case for a large truck and a small car, *object overlap* is the stronger or dominant visual cue to distance.

Another visual cue to distance in the visual field, especially at distances, in tasks like driving, is the texture-density of the field in the foreground of the object. To be effective as a cue to distance, the texture-density has to have regularity or consistency as might be found in an asphalt-covered roadway. The density of elements increasing continuously with distance is a strong cue used by the observer to judge both the distance of an object from the observer and the relative distance between two objects in the visual field. Note that the elements within the texture field have to be of sufficient size and contrast so that the cue is visible in the field under prevailing light conditions.

It is often the case that multiple cues to distance are available in the same visual field. Texture-density cues in a roadway will likely be accompanied by linear perspective cues. Linear perspective cues occur when the observer views parallel lines extending into the distance. Optically, the lines appear to merge and the merge increases in intensity until they appear to touch. Linear perspective cues to distance are an artifact of human cultural activities such as roads and buildings and rarely occur naturally.

Even on level roads and highways, an object further away will appear higher in the visual field than closer objects. This *height in the visual field* cue is yet another strong visual cue to distance. An object perceived as higher in the visual field will always appear further away than objects lower in the field even if they are unfamiliar to the observer.

Two other cues to distance are applicable to objects much closer to an observer. Both are integral to the operation of the eyes, independent of object characteristics, and are purely reflexive in nature. *Convergence* refers to the fact that both eyes tend to converge on an object which is the focus of visual attention. This convergence begins at about 50 cm (20 in). The feedback from the small ciliary muscles in the eye, along with the increase in the object's perceived size, conveys the observer that an object is coming very close to the observer's face. The strength of this cue increases linearly in intensity as it approaches. As the object moves away from the observer, the eyes diverge accordingly.

A second cue to distance is the change in the shape of the lens of the eye in order to accommodate object distance. *Accommodation* occurs reflexively in the eye in response to image blur. Accommodation begins as early as 10 m (about 33 ft) but is strongest at distances of under 2 m.

Both convergence and accommodation are likely to be strong cues only at distances much less than the other visual distance cues discussed above. The cues will likely play a role in judging distances within the vehicle rather than outside. Part of this is due to the limitations of the cues' physical parameters. It is also partly due to the presence of other cues to distance that occur at distances outside the vehicle.

Collectively, visual cues to distance can be considered to be pre-attentive as the observer does not need to be consciously aware of their effect on distance or depth judgment. Second, the use of more than one of these cues tends to be additive in their effects on distance judgments. Thus, the more such cues are available the more accurate the distance judgments are likely to be. An observer's judgment of distance is, therefore, likely to be more accurate if the observer scans the area for additional visual cues than to rely solely on one cue such as the optical size of an object.

NON-VISUAL SYSTEMS

While it is understandable to consider visual perception as predominant in driving, the driver is equipped with other senses that are often overlooked. Among these are audition (a sense of hearing), vestibular (a sense

of physical motion), tactile (a sense of touch and pressure), and proprioception (a sense of the position of limbs). Each of these cues contributes
important information to the driver.

AUDITION

A sense of hearing is often taken for granted in daily tasks like driving
until, due to age or other factors, it no longer functions properly. Hearing is
essential in verbal communications, which occur with amplitudes between
40 and 60 dB. The frequency range of communications is gender-depen-
dent but ranges from 85 and 255 Hz. The full range of receptive hearing is
between 20 and 20,000 Hz although the higher frequencies are lost with
increasing age, declining to 12,000 Hz after age 50.

Hearing supports driving tasks in a variety of ways. Auditory cues
such as engine RPM and wind noise are cues to control vehicle speed. In
vehicles with manual transmissions, engine RPM amplitude and frequency
changes signal the driver to change gears. Unusual sounds emanating from
outside the vehicle can portend vehicle component failures, while others
can indicate an imminent departure from the road as in the case of the
sounds made from tires hitting "rumble strips". The placement of an ear
on each side of the head provides the hearer with the ability to locate the
source of sounds by relying on the time and amplitude differences between
the two ears of the associated sound wave. This function of hearing is
essential in identifying the location of oncoming emergency vehicles. The
amplitude and frequencies of these sounds are deliberately designed to be
heard within the vehicle and to overwhelm any other sounds that the driver
may be attending.

Apart from sounds emanating from outside the vehicle, the driver relies
on a variety of sounds from within the vehicle. These sounds include, but
are not limited to, auditory feedback from the operation of turn signals
and from the use of touch interfaces such as those found on information
and entertainment displays. Increasingly, newer in-vehicle technologies
such as navigational aids provide verbal as well as visual instructions. It
is expected that these technologies will likely require verbal instructions
as well as auditory alerts and alarms in an attempt to reduce driver visual
attention to displays and controls inside the vehicle.

SOMATOSENSORY SYSTEM

The somatosensory system or body senses include the tactile sense or
sense of touch, a sense of where body limbs are in space, and a sense that
responds to the physical movement of the body. The tactile sensations of a
driver might seem rather trivial at first glance. However, the sense of touch,

specifically the sensation of pressure, plays a vital role in the operation of vehicle controls and increasingly that of displays. Tactile sensation is provided by mechanoreceptors in the skin. The receptors respond to pressure, stretching, temperature, and other stimuli. These receptors are at a very high level of density in the fingertips. The high density allows the fine sense of touch needed for the manipulation of even the smallest controls. Mechanoreceptors are also found in the palm of the hand, which allows the driver to exercise accurate grip strength. Similarly, the mechanoreceptors in the base of the feet allow for the sensing of pedal placement and pressure.

Mechanoreceptors are also in the skin of the torso and seat though at lower density levels. These provide feedback from the sensation of vehicle vibration and physical forces acting on the body as would occur during vehicle accelerations and decelerations and from those forces that result from driving around curves.

Considered another element in the somatosensory system along with touch, proprioception is the perception of the position of a body limb. The proprioceptive sense is generated by components within the musculoskeletal system. These components send signals when muscles or joints are activated. Proprioception is a valuable adjunct to vision in the location and operation of controls but is essential to the use of controls not normally viewable to the driver. This is the case for all foot controls such as the accelerator, brake, and clutch pedals. This is also true for controls requiring hand movements such as gear shifting and emergency braking levers. As a driver becomes more familiar with the controls of a vehicle, a proprioceptive sense can aid in the operation of other controls of a vehicle such as turn signals and other controls mounted on the steering wheel. The development of this proprioceptive sense allows the operation of the control without dependence on vision, an important sensory-perceptual resource. This is one of many cases where resource management through skill development can reduce dependence on a limited sensory resource like vision. A more detailed discussion will be provided in Chapter 7.

Vestibular System

In addition to the somatosensory system, other organs in the body are uniquely designed for assessing changes in the physical forces that drivers experience. For sensing linear accelerations and decelerations, the otolith within the vestibular organ can sense changes in the velocity of 8.5 cm/sec^2 in fore-aft movement and 6.5 cm/sec^2 in lateral motion (Kingma, 2005). Physical movement in rotation also plays a role in driving. The semicircular canals, also part of the vestibular system, sense changes in movement velocity in rotation as low as 0.11 deg/sec^2 (Clarke and Stewart, 1969). For the driver, both linear and angular forces may occur simultaneously such as in high-speed turns. At higher rates of acceleration and deceleration, these forces not only trigger vestibular sensations but also provide pressure sensations to the torso and seat.

PERCEPTION-ACTION SEQUENCES

In driving, as in many activities, certain perceptual functions can become coupled to an action sequence such as a grasping action to a door knob. In this example, the affordance visual cue of the round knob is coupled with the physical grasping behavior. Perception-action sequences can occur as a result of directed training or formed by repeated experiences. In driving, a variety of both simple and complex perception-action sequences are triggered usually by visual stimuli either internal or external to the vehicle. The responses to these stimuli are action sequences typically composed of individual motor responses that involve specific learned patterns of action. An emergency braking response, for example, is triggered when the driver perceives an immediate collision hazard where space or time may not allow for maneuvering. The braking response involves rapid movement from the accelerator to the brake pedal followed by maximum pressure on the brake pedal itself. With repeated practice the emergency braking perception-action sequence may become overlearned. An action sequence is overlearned when the driver continues to repeat the response long after complete learning has occurred. This overlearning leads to *automatization* of the sequence. Automatization is achieved when the action sequence no longer requires the driver to focus attention on the behavior. Instead, the behavior is automatically executed when the specific perceptual trigger is processed by the driver. Automatized sequences can be contrasted with controlled sequences where the latter still require the driver to focus attention on the sequence in order to complete it (Shiffrin and Schneider, 1984).

The most important class of perception-action sequence is perceptual-motor in nature. That is, the action sequence always involves a physical action of some sort. Visuomotor sequences are a type of perception-action sequence. They typically pair a visual stimulus with a physical action as in the emergency braking example. Visuomotor sequences can involve a "closed loop" where a changing visual stimulus initiates a physical response. In the case of steering, visual stimuli from the road surface are used as guidance to maintain lane position. Small changes in perceived lane position will result in steering wheel inputs to maintain lane position. The need to update lane position becomes more frequent when the road curves require more precise adjustments in response to changes in the curve tangent (see Chapter 2).

More complex perceptual-motor sequences occur in driving than in either braking or steering. The manual transmission, once the most common means of changing a vehicle's gears, is increasingly being replaced by automation. However, those vehicles that still require manual gear shifting are examples of complex visuomotor action sequences that become automated with experience. The action sequence is typically initiated by the driver's use of a visual indicator such as a tachometer or even engine sound to determine the time to change gears. The action sequence is then initiated beginning with pressure on the clutch pedal, movement of the gear shift level to the desired gear, and then release of the clutch pedal. The procedure is executed automatically and expeditiously to avoid as much speed loss as possible. Note that while a visual or auditory stimulus initiates the action sequence, the sequence itself is executed with only proprioceptive and tactile feedback.

A variety of perceptual-motor action sequences exist in driving vehicles. They vary with the vehicle type and how it is equipped. While the rudimentary tasks of braking and steering are found in all vehicles, an increasing number of control and display features are entering the vehicle cab. These systems will demand resources of the driver, not only visual resources but cognitive resources as well. Automatizing the action sequences required by these systems may be one way in which the demand for driver resources may be reduced.

COGNITION

Perceptual processes including perception-action sequences begin with an external stimulus. Perception, including all the types discussed thus far, is stimulus-driven, and the resulting percept such as distance or depth is dependent on this external energy source. Cognitive processes can be initiated by either an external stimulus or an internal event within the brain. Cognitive processes include storage and retrieval from working and long-term memory. These processes are prevalent in the process of developing situation awareness in driving. Wayfinding uses these processes to store and retrieve environmental landmarks. These and their spatial relationship are used in the construction of cognitive maps stored in the driver's long-term memory. Hazard perception also depends on memory processes to recognize and classify hazards to the vehicle including collision and loss of control.

SELECTIVE ATTENTION

Selective attention is a cognitive process that focuses the physical and mental resources of the driver toward a particular stimulus, event, or task.

These may occur either within or outside of the vehicle. Selective attention is one of the most important cognitive processes for operating a vehicle, and perhaps the most important tool a driver has in managing limited resources. It is, however, a tool that requires mental effort (Kahneman, 1973) which means its use comes with a significant cost to the processing capabilities of a driver.

Selective attention is involved in controlling vision, audition, and other sensory processes. In applying visual senses, vision is categorized as either focal or ambient and pre-attentive. In focal attention, visual resources, specifically those in central vision, are focused on the particular stimulus for the purpose of processing the stimulus further. It is a consciously controlled process that prioritizes processing to the exclusion of other visual stimuli. The expenditure of focal vision and the time and mental effort of focusing attention make the use of focal attention expensive with regard to the expenditure of resources. For this reason, focal attention in vision or in any other senses should be reserved for tasks of a high value to the driver in performing driving tasks.

Those stimuli in the periphery of vision are considered to be subject to ambient awareness but not to selective attention as such. The most obvious visual stimuli subject to ambient awareness is the optic flow that informs the driver of the speed of the vehicle and its lane position. The driver does not need to focus attention on the optic flow for it to affect awareness of speed. That is, the driver is aware of these stimuli but does not need to actually attend to them. It might be said that ambient stimuli are those that are in the background of experience and not in the center of it. Provided the ambient stimuli are of lower importance to the driver in the immediate task, this low-level awareness is an example of an efficient use of sensory resources.

Peripheral vision, due to its sensitivity to movement, is particularly sensitive to the movement of objects. Such moving objects can be hazardous to the driver's vehicle, and detecting them with an otherwise unattended resource is another important resource management function even though the driver is not directly controlling it. Ambient awareness through the visual periphery is available under both photopic and mesopic driving conditions. While the driver responds to peripheral motion cues, visual processing of light stimuli in the visual periphery is particularly strong in mesopic conditions where the contrast between the light stimuli and its background is significant.

The focal and ambient dichotomy applies to other sensory systems than visual. Auditory selective attention is often used to focus on verbal communications, which is considered to be more important to the listener than other sounds, including verbal communications as illustrated by the "cocktail party" phenomena. In this case, focal attention is on a communication of particular interest while others are left to a lower-level ambient awareness. For drivers, selective focal attention to non-driving tasks such as cell phone use may leave important driving information to only ambient awareness as in the case of verbal warnings from a passenger regarding a road hazard.

Selective attention is not always voluntary. Initiation of the process may be the result of a high-intensity stimulus event such as bright lights or loud sounds. The stimulus is of such intensity that it demands a shift of attention away from whatever task is at hand. While voluntary selective attention involves a deliberate conscious act toward a particular task, involuntary attention shifts are the result of external, unique stimuli typically of high intensity. While voluntary shifts in attention allow an orienting response that allows for some anticipatory preparation time, the involuntary shift in attention does not. This lack of preparation time results in a higher likelihood of responding incorrectly to the new stimulus event than for responses to voluntary shifts. Responses to visual or auditory alarms, for example, may be successful in generating a startle response but not necessarily a response appropriate to the cause that triggered the alarm.

SUSTAINED SELECTIVE ATTENTION

A common question in driving is just how long the focus of attention on a given task can be maintained in the presence of other secondary or distracting tasks. Sustained selective attention, although related to vigilance (below), may be described in terms of an individual's ability to concentrate over time on a given task regardless of the level of effort the task demands. The hazard events are relatively rare, thanks to the design of the road system itself. The level of concentration needed for a particular task increases not only with the task but also with the number of competing secondary and distracting tasks. A more detailed discussion of this subject for drivers will occur in Chapter 7.

VIGILANCE

Vigilance refers to the general state of the individual's alertness rather than the level of focused attention needed to perform a specific task. The vigilance of a driver or the overall alertness of the driver is affected by such factors as fatigue, sleep deprivation, and the individual's overall health. These impact tasks involve selective attention, but they may also affect other tasks that do not require selective attention but do need highly functioning physical resources.

WORKING AND LONG-TERM MEMORY

Memorial processes are subject to failure during either the storage or retrieval point due to a variety of issues. In situation awareness, the limits of working memory itself restrict the number of elements that can be stored. For example, the number of potential vehicle and pedestrian hazards at an intersection may be too many for the driver to maintain in working memory. Typical numbers for working memory limits are three to four chunks of information, where a chunk is a grouping of elements that occur naturally or by a deliberate action of grouping by the individual. In driving, this grouping of elements might be difficult for a driver, particularly if position and threat levels have to be included.

In theory, the capacity to store information in long-term memory is unlimited. Failures in retrieval from memory are, however, all too common. This occurs in wayfinding when the landmarks have similar characteristics. As a result, the driver may make the wrong turn at a landmark that is mistaken for another stored within the cognitive map. Failures in hazard recognition may occur due to an inadequate representation in long-term memory. This may occur in recognizing attributes of behavior that a pedestrian is exhibiting. Inexperienced drivers often fail to recognize that a pedestrian approaching the roadway crosswalk may continue into the roadway regardless of the status of the traffic signal. The driver's memory does not contain a complete representation of the pedestrian hazard and makes a decision based on incomplete information.

Wayfinding and navigation through the driving environment is a routine task for drivers. In general, most driving is conducted in familiar environments, along familiar routes, with familiar landmarks. The landmarks serve as key points in the construction of an internalized cognitive map built by the driver through repeated exposures to the same route. The cognitive map or portions of it are retrieved from long-term memory to provide the driver with needed location reference points and guidance for turns. The maps can be of theoretically unlimited size but are subject to retrieval errors including interference from other driving tasks. Wayfinding will be discussed in more detail in Chapter 5.

Long-term storage and retrieval are particularly relevant in the identification and classification of potential hazards. Hazards are classified as such because they contain or are associated with specific attributes which define an object or a situation as a threat. Each time a driver visually scans the road ahead, the attributes stored in the internal representation are matched to those in the visual scan. If enough characteristics are present, the classification criteria of a potential hazard are met. This classification process takes time so visual fixations, not just scans, are required. The original storage of these representations should occur early in driver training and be reinforced with experience. Many drivers do not receive hazard perception training, and others are only exposed to a select few hazards and limited exposure to different driving environments. A detailed discussion of hazard perception is provided in Chapter 4.

The development and retrieval of internal representation occurs in driving other than cognitive maps and hazard perception. Internal representations of the vehicle's external frame or spatial footprint are also likely to be developed as the driver's experience with the specific vehicle increases over time. This internal representation is especially useful in proximity judgments inherent in tasks such as parking or when the vehicle is in close proximity to other vehicles or pedestrians.

SUMMARY

This chapter highlighted a broad range of sensory, perceptual, and cognitive processes involved in driving. The visual and non-visual processes examined reveal the capability and limitations of sensory input from within and outside the vehicle. The limited high-resolution visual field demands that the driver scan and search the road environment actively rather than passively awaiting events. The visual feedback from the optic flow helps the driver estimate speed without reference to the speedometer. Visual guidance can be derived from the changes in the direction of that same flow field. While visual acuity (VA) plays some role in driving, particularly in small object hazard detection and critical details of certain objects (e.g., road signs), visual contrast sensitivity is likely more important in night (mesopic) driving. Factors affecting distance perception such as the change in object retinal size and texture-density are important elements in a variety of driver tasks. Non-visual sensations including hearing, touch, and responses to the physical forces that vehicle driving imposes are discussed. Finally, cognitive processes, specifically selective attention and working memory and their importance to driving and to the development of a situation, are described.

NOTE

1. The term "photopic" will be used instead of daylight to avoid any ambiguities.

2 Vehicle Speed Control and Guidance

Two essential components of driving are the control of vehicle speed and the ability to control the path of the vehicle under a variety of conditions. In this chapter, driving speed and guidance will be discussed in terms of three major components: driver perceptual and perceptual-motor capabilities, the road environment including lighting and weather, and the vehicle design. These three areas interact to affect driver performance in a variety of ways. In order to determine the effect of driver behavior on the speed and guidance of the vehicle, it is necessary to decompose these two components into their respective perceptual and perceptual-motor constituents to reveal the contribution of each in the processing of incoming information. Speed and guidance information is primarily visual, but non-visual information is also available through the auditory, vestibular, and somatosensory systems.

VISUAL PERCEPTION OF VEHICLE SPEED

In the previous chapter, the role of information from the optic array was discussed in some detail as a means of controlling both self-motion and the direction of movement. As the observer moves through the environment, an optic flow is created. This optic flow occurs during photopic and mesopic movement. The optic flow intensity and therefore the intensity of self-motion are dependent on a variety of factors. In naturalistic settings, such as those found in driving, the optic flow intensity can vary widely from one situation to the next. Regardless of the driving situation, the ability of the driver to estimate vehicle speed is essential in order to avoid an over-dependence on the vehicle's speedometer. Excessive time spent viewing the speedometer means that much less time is spent extracting guidance information from the road and searching for road hazards.

The ability of drivers to estimate vehicle speed has been the subject of a number of studies. Among the earliest of these was conducted by Recarte and Nunes (1996). This study was conducted on a test track consisting of straight and curved road segments under photopic conditions and clear weather. For all tests, the speedometer of the vehicle was covered. The drivers were required to match three target speeds: 25 mph (40 kph), 50 mph (80 kph), and 62 mph (100 kph).

The data collected reveal a consistent *underestimation* of the target speed at all speeds tested. A more recent research simulator study by

DOI: 10.1201/9781003454373-2

Durkee and Ward (2011) replicates the findings of the Recarte and Nunes (1996) study. Both of these studies show the drivers' tendency to underestimate their speed and, as a result, to drive faster than the target speed. An important aspect of this speed underestimation is that it is reduced to an average of 107% at speeds above 60 mph from an average of 134% at the lowest speed of 25 mph across the two studies. The propensity to drive much faster than the target speed in the low (25 mph) speed regime is of particular concern as this lower speed is typically associated with urban environments which typically have a higher number of road hazards per mile than roads with higher speeds.

An additional field study of driver speed judgments showed similar results. In this study, passengers judged vehicle speed passively while drivers attempted to produce the desired target speed (Schutz, Billino, Bodrogi et al., 2015). In both cases, speed was underestimated by about 20% compared to 34% for the previous studies. Moreover, the underestimation of speed was about the same in both mesopic and photopic conditions. This study also showed the pattern of increased underestimation of vehicle speed at the lower speed of 25 mph (40 kph) when compared to the higher speed of 50 mph (80 kph). Notably, this study was conducted on an abandoned runway without surrounding objects and with only center, discontinuous road markings.

As none of the above studies allowed the drivers to view the speedometer, information regarding vehicle speed needed to be extracted from either visual or non-visual sources of vehicle speed. As vision is primary in the operation of vehicles, the most likely source of speed is the optic array, that is the entirety of visual stimuli available to a driver's retinae at any one time. The optic array and the perception of speed in humans were discussed earlier in Chapter 1. However, it is necessary to decompose that array for the specific task of driving in order to determine the relative contribution of components to the issue of vehicle speed perception.

Optic Flow

Anytime an observer moves through space, an optic flow is created that corresponds to the speed of the observer. The perception of that speed is not, however, a direct correlation with the optic flow component of the optic array. Instead, the perceived speed is a function of the stimulation of the observer's retinae by perceptible elements in the array. That is, the size and relative contrast of these elements must be above the visual threshold for the position of that element in the observer's visual field.

In driving, the constituency of the optic flow field is dependent on a number of factors. The first of these is the functioning of the driver's visual system including the visual periphery and macular or central visual fields. The performance of the driver's visual system under both photopic and mesopic conditions affects the ability to process optic flow efficiently. The perception of vehicle speed will be affected by any loss of visual function. As the visual periphery is particularly sensitive to movement, problems with peripheral vision are likely to have the largest effect on speed perception.

Eye Height

Assuming a normally functioning visual system, driver eye height above the road surface is also a known factor affecting the perception of speed. The optic flow rate is scaled by driver eye height above ground (Warren, 1982). This is due to changes in optic flow field characteristics, specifically the angular velocities of flow elements. These elements, particularly those close to the driver, decrease in speed as eye height is increased. This results in driver perception of slower speed when compared to driver eye heights closer to the ground. Driver eye heights vary depending on the class of vehicle driven, from common passenger cars at around 1.08 m to large trucks at nearly 3 m above ground level. The strongest optic flow will be experienced by those driving sports cars with a typical eye height of about 0.8 m. (39 in). The corresponding reduction in perceived speed with increased eye height can lead to unsafe levels of speed in some vehicles. The high incidence of rollovers in Sport Utility Vehicles (SUVs) may be due, in part, to the higher eye height compared to other passenger cars as a result of the SUV's high center of gravity. Increasing driver eye height in an SUV driving simulator has been found to result in higher driving speeds than when eye height was lowered (Rudin-Brown, 2004). Driver eye height is also known to affect vehicle speed production in large trucks. An increase of only 33.2 cm (13 in) in the driver height resulted in a 12% increase in speed when compared to a smaller truck model under the same conditions (Panerai, Droulez, Kelada et al., 2001).

Road Environment

The second component affecting the strength of optic flow for drivers is the road environment itself specifically the presence of objects and their location with reference to the driver's visual field. These include vehicles and buildings, road signage, guard rails, road texturing, trees and foliage, and other objects. In the case of driving in tunnels, the texturing, patterning, and other attributes of the tunnel walls may affect the optic flow and the perceived vehicle speed that results. Note that the extent of the flow field experienced by the driver will be constrained by the degree to which the vehicle structure reduces the FOV available to the driver at any time. For example, the flow field for a motorcycle is largely unencumbered by the vehicle frame, which results in a strong flow field and a heightened perception of vehicle speed.

The visual clutter of the road environment has been shown to affect a driver's perception of speed by increasing the strength of the flow field. The study by Durkee and Ward (2011) described above found that the addition of visual clutter to the roadside improved driver accuracy in matching target speed when compared to a condition in which visual clutter was low. Moreover, road surface texturing of areas adjacent to a road has been shown to affect driver speed when compared to a condition when no texturing was available (Kountouriotos et al., 2016).

The precise location of road components most likely to impact the perception of speed is related to where the highest angular velocities of the flow field are for the driver. These tend to be in the far periphery of the driver's horizontal field as much of the vertical field is blocked by the vehicle structures. Image display characteristics on driver speed perception that have been investigated in driving simulators can be informative on this issue. Jamson (2000, 2001) increased the horizontal field of a simulator from 50 to 120 deg and finally to 230 deg. No increase in driver speed perception was found beyond 120 deg horizontal FOV. Note that a 120 deg horizontal field extends well beyond the driver's FOV of the windshield width of a full-sized passenger car. Thus, driver speed perception is likely to be influenced by the flow field perceived through the driver-side window as well as a portion of the windshield near the vehicle's A pillar.

Edge Rate

A second component of an optic array is *edge rate*, which is also known to affect the perception of speed. Edge rate is the rate at which a visual field element crosses a visual location which is fixed relative to the observer. In driving, this fixed position is usually some portion of the windshield, particularly the lower edge. Laboratory studies show that optic flow and edge rate are additives in their impact on speed perception, with edge rate accounting for a larger proportion of the overall effect (Larish and Flach, 1990).

The most obvious edge rate effect for driving occurs from the discontinuous lane markings used to separate lanes. Other sources of edge rate information include the support poles for guard rails, markings on jersey barriers used to separate opposing traffic on multilane roadways, and expansion gaps on bridges, to name just a few. The central feature that makes edge rate elements of value in estimating speed is their regularity in the separation from one edge element to another. A second needed attribute is that the edges need to be present for extended periods of time.

Discontinuous lane separation markings are the most common instances of edge rate information. On main U.S. highways such as interstate freeways and expressways, the standard length of the marking is 3.05 m (10 ft) separated by a gap of 91 m (30 ft). For roadways traveled less frequently or at lower speeds, different combinations of marking length and separation are used. The ability to judge the actual length and separation gap by drivers is quite poor. A study of college-aged drivers found estimates of both the length and separation of standard markings to be the same at about 0.61 m (2 ft) (Shaffer et al., 2008). This is likely due to the perceived visual compression of both resulting from the angle of incidence between the driver and the road surface.

Due to their known contribution to perceived speed, edge rate variations and their impact on driver behavior have been studied extensively. Thus, a spacing of regularly appearing road patterns can easily be manipulated to alter the perceived temporal frequency of edges. Denton (1986) systemically manipulated the temporal frequency of edges and found that low frequency resulted in consistent driver underestimation of speed; i.e., the driver drove the vehicle faster than the intended target speed while the reverse held true for high temporal frequency edges. Similar effects have been found in other studies. A study conducted in a driving simulator altered the temporal frequency of centerline markings and found increases in perceived speed with increases in frequency (Pritchard and Hammett, 2012).

More aggressive means of applying the effect of edge rate on perceived speed have been investigated. Application of transverse lines of varying dimensions to the road surface has been applied in an attempt to slow driver speed prior to entry into curves. (Excessive speed is a known cause of accidents in curves.) Transverse bars of decreasing spacing were applied to rural road curves with a speed limit of 62 mph (100 kph) (Comte and Jamson, 2000). A similar study used transverse lines extending 0.6 m from the lane edge to also be effective in reducing driver speed (Godley et al., 2000). The transverse markings were effective in reducing driver speed prior to curve entry. Hatched center lane markings have also been found to reduce driver speed (Godley et al., 2004).

A final example of applying the edge rate effect to driver speed control has been in studies of driving speed in tunnels. It was noted earlier that the tunnel wall markings may serve as edge rate information provided they were regular in frequency and occurred over long distances. Too often, tunnel walls provide very little in either optic flow or edge rate information due to their construction. This leads to higher driving speeds in tunnels than would occur in other road environments where the optic array was more informative with respect to motion. Tunnel wall markings as a countermeasure to driver speeding behavior are made more difficult by the greater distance from the driver and the poorer light levels normally available when compared to the open highway or street. Experiments in sidewall markings of tunnels have manipulated both temporal frequency (0.4 to 32 Hz) and color contrast (red, yellow, and blue each paired with white) (Wan et al., 2016). The red-white color contrast had the greatest effect of increasing driver speed overestimation and with it reduced vehicle speed. Temporal frequencies between 4.45 and 7.01 Hz appeared to have the strongest effect in reducing driver speed in tunnels. In a more recent study of tunnel markings, temporal frequencies of 2 to 32 Hz produced the most driver overestimation of speed, while markings less than 1.0 Hz tended to produce driver underestimation of speed (Zeng et al., 2018).

LUMINANCE

It should not be surprising that a system such as vision which is so dependent on light for its function would be affected by the reduction of light energy that comes with some driving conditions. The first of these problems comes when the available light energy drops below 3 cd/m², that is, only mesopic, rather than the photopic, energy that is available. As mesopic visual function reduces the driver's contrast sensitivity and visual cue processing speed, the effects on low luminance speed perception are that driver speed perception worsens under night driving conditions even with the use of headlights. The driver will suffer from both poorer visual acuity for high value and focal tasks, as well as from poorer performance in ambient tasks such as object motion detection due to the slower response speed of rod receptors in mesopic vision (Gegenfurtner et al., 1999).

In laboratory studies of the effects of luminance on the perceived speed of moving gratings, the influence of luminance depends on the speed of movement. Thus, at slow speeds of less than 4 deg/sec, mesopic or photopic luminance levels had no effect on perceived speed. However, at speeds above 4 deg/sec, the gratings appear to be low luminance moving faster under mesopic luminance when compared to photopic conditions (Hammett et al., 2007). In a later study by Vaziri-Pashkan and Cavanagh (2008), subjects matching rotating patterns of gratings overestimated the speed of low luminance gratings moving under 4 deg/sec. The overestimation increased with increased speed up to 10 deg/sec of motion. The findings reflect the phenomenon of motion smearing or blurring at low luminance levels likely due to the slow response speed of rod receptors.

The differential effect of luminance levels has implications for perceptual countermeasures in the use of road markings for speed control as well as the tendency of drivers to drive faster at night than during the day under the same road conditions. Under mesopic conditions, the poorer response speed of rod receptors and their inhibitory effect on cone receptors of the eye affects the driver's ability to judge speed accurately. The reduced sensitivity of the eye to movement under mesopic luminance levels results in a blurring or smearing of the resulting image. This has the effect of reducing the salience of optic flow patterns, which are most effective when individual elements are clearly perceptible. Smearing or blurring due to low luminance reduces their influence and thus their effect on perceived speed. The previously mentioned study by Pritchard and Hammett (2012) found that the increase in perceived speed with increased edge rate occurred under both mesopic (0.4 cd/m²) and low photopic (5 cd/m²) driving conditions. However, driver speed was reduced with the lower luminance in all edge rate conditions. Employing countermeasures that have been described above needs to consider their effectiveness under the low luminance levels found in mesopic driving conditions where image blurring is common.

Lower luminance effects on the visual system have implications for road geometry design and countermeasures. Drivers tend to enter curves under night conditions at speeds much higher than those during daytime conditions due to poorer speed cues (Pretto and Mangaro, 2010). Additionally, drivers have difficulty discerning road geometry both during the daytime and especially at night. This results in entering the curve at high speeds where drivers may be unable to decelerate in time to negotiate the curve.

Contrast

Reducing field contrast effects uniformly also results in lower visual stimulation emanating from optic flow and edge rate elements. Again, this results in the reduced perception of speed at slower vehicle speeds resulting in the driver increasing vehicle speeds to meet the target. This has been demonstrated in a variety of studies on uniform contrast reduction (Stone and Thompson, 1992; Owens et al., 2010). This does not appear to apply to non-uniform or exponential changes due to, for example, fog (Pretto et al., 2014). In this case, the optic flow field's higher angular velocities are still available at the far visual periphery of the driver even though the central field visibility is reduced. The driver's reduction in speed in fog is not due to the optic flow field but to the increased perceived risk of collision (see Chapter 4).

Edge rate effects of road markings will also be improved with increased contrast between markings and the road surface. Improving reflectivity, such as the application of advanced retro reflectivity paints to discontinuous centerline markings (3 m stripes followed by 9 m gaps), has been shown to affect speed control at night (Horberry et al., 2006). Both target speed compliance and reduced speed variability at a target speed of 62 mph (100kph) were found in this study.

NON-VISUAL PERCEPTION OF SPEED

The importance of the driver's visual system, the road environment, and elements of the vehicle itself has been discussed with reference to the perception of speed. The role of non-visual sensory and perceptual systems needs to be addressed as well. Three of these systems play a role in the control of speed by a driver: auditory, vestibular, and somatosensory systems.

Auditory

Auditory feedback from vehicle tires and wind effects on the vehicle frame and engine contribute to the perception of vehicle speed. In the absence of these sounds, target speeds in the 30 and 70 mph regimes have been shown to be harder to maintain with the speed control error greater at 70 mph (Merat and Jamson, 2011). At speeds of 30 mph or less, acoustic feedback from the vehicle is likely to be much less, particularly on well-maintained roads in urban areas. With increases in aerodynamic efficiency and construction, modern passenger cars have substantially reduced wind and other noises from entering the vehicle. This means that the driver when not attending to the speedometer is even more heavily dependent on vision for vehicle speed control.

Vestibular

The effects of accelerating and decelerating a vehicle are sensed by the driver's vestibular system if they exceed the sensory threshold for longitudinal, linear changes in velocity. The threshold has been estimated to be about an order of magnitude higher in vehicle operations than the laboratory-measured sensory threshold mentioned in Chapter 1. Angular accelerations as experienced when the driver's head is rotated are also experienced in driving but are often accompanied by linear accelerations at the same time. For example, in negotiating curves, drivers may experience physical motion as a lateral acceleration if the curvature is relatively high. This will occur with greater intensity if the speed entering the curve is excessive. However, because the vestibular system is a mechanical system dependent on fluid movement, it is much slower to respond than other sensory systems. For this reason, the lateral acceleration that imparts to the driver the sense of departing the roadway usually occurs too late for any corrective action that might be needed.

Somatosensory System

The somatosensory system of the driver responds to pressures on the body surfaces that occur when the vehicle is in motion. These include the vibrotactile sensations resulting from the vehicle's tires moving across uneven road surfaces. The higher the vehicle speed, the greater the vibration experienced by the driver. Drivers tend to dislike high levels of sustained vibrations over long periods of time and will avoid roads with rough surfaces.

As with changes in vehicle velocity sensed by vestibular mechanisms, somatosensory body cues such as pressure cues from the seatbelt and torso movements accompany linear and angular changes as well. As the somatosensory system is more sensitive than the vestibular system, these body cues are likely to be more efficient in informing the driver of changes in vehicle state.

VISUAL PERCEPTION IN STEERING CONTROL

Coincident with the task of controlling the speed of a vehicle, that is its linear movement, is the need to control its lateral movement, particularly in curves. The means to do this is through steering wheel inputs, and the visual input necessary to accomplish this task consists of two information sources. The first source or component supports what is in effect a compensatory tracking task which combines the visual input extracted from the road surface with steering wheel inputs. The visual input area of the road surface is only about 1 to 2 sec ahead of the vehicle. The combined visual and manual inputs form a closed loop in which updates to changes in the visual path are matched by updates in steering wheel inputs. A second component supports the control task by providing lead or anticipatory inputs from an area about 2 to 3 sec ahead of the vehicle (Donges, 1978; Salvucci and Gray, 2004).

This so-called "two-component" model of steering has been experimentally verified by studies restricting the forward view of the driver by blocking portions of the driver's forward view in research simulators. Only a small portion of the visual field, as small as 1 deg vertical, is needed for each near or *compensatory view* and for each far or *anticipatory view* (Land and Lee, 1994). The anticipatory view is necessary for smooth steering control. When anticipatory information is blocked from the driver's view, the steering control becomes more erratic. This control behavior has been described as "bang-bang" control as the driver attempts to respond to information from the near view ahead of the vehicle. The lack of time to anticipate future path changes results in reactive rather than proactive control behavior.

OPTIC FLOW

One might ask at this point what precisely is the driver processing from the visual field ahead that supports steering control. One of the likely candidates is the element of the optic array in the optic flow field. The optic flow field in driving is determined by the heading of the vehicle. As drivers are generally looking where they wish to go, the optic flow field or some component of it would seem to be a logical candidate for steering control.

As noted earlier, the optic flow field includes all of the perceptible elements flowing past the two retinae. An important issue with the use of the optic flow field for guidance is to determine which of the elements of the flow field is likely to be most useful. Heading information can only be derived from the flow line locomotion that flows beneath the observer (Lee and Lishman, 1977). Computer simulations have shown that gaze direction on curvilinear paths can only support guidance when the velocity vectors become linearized (Kim and Turvey, 1999). In other words, the driver's eyes must be moved until flow field velocity vectors are aligned with their gaze. At this point, the driver is looking in the direction they want to travel. The optic flow field vectors now become useful for guidance, and steering inputs can provide the necessary compensatory tracking to maintain alignment between the vehicle and the intended path of travel. As long as the driver's gaze is aligned with the direction of travel as defined by the linear elements of the optic flow, the flow field is useful for guidance. The optic flow field, specifically the linear elements, can provide guidance in the absence of delineated pathways such as roadways. Off-road vehicle guidance or even guidance on roadways contaminated by snow, ice, or other elements can be provided by optic flow, provided some linearized elements are perceptible.

DELINEATED PATHS

For most driving, delineated paths in the form of roadways simplify the guidance problem. They provide both road edge and lane position markings, which clearly delineate the path that must be followed. In maneuvering curves, eye movement data show that the driver is focusing on the tangent of the curve for guidance when no vehicle traffic is ahead (Land, 1998; Shinar et al., 1977). The tangent of the curve is the angle formed between the vehicle heading and the road edge when turning right or the road centerline or lane separator if turning left (see Figure 2.1). The angle is maintained throughout the negotiation of the curve. This angle is held constant throughout the curve where the maximum curvature of the road is expressed. Land (2001) argued that the road edge of the curve provided the essential cue for steering. If road markings were poor or absent, then road texturing in and around the road edge could be used as an adequate substitute. This is supported by the presence of texture cues in improving lane position accuracy (Chatziastros et al., 1999). However, road edge markings due to their perceptual saliency represent the more important cues to steering, particularly the interior edge lines of the curve. Clearly defined and high-contrast road markings both on the road edge and for lane separation are important for steering control.

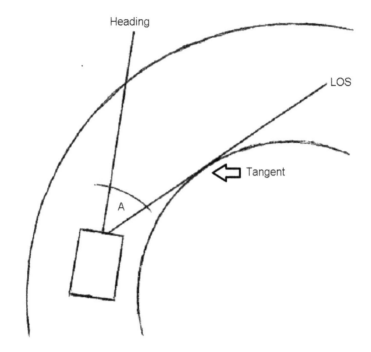

FIGURE 2.1 Tangent of the curve with bearing angle (A) and line of sight (LOS).

The demarcated or delineated path enhanced by road markings is essential for accurate steering control in driving. However, when road edge information and optic flow variables are manipulated experimentally, optic flow was to modulate the influence of road edge information (Mule et al., 2016). An additional study by Kountouriotis et al. (2016) found that while the delineated path (road) provides sufficient information to steer, the optic flow field does not. Moreover, the manipulation of road edge lines in curve driving shows a strong effect on steering control. In a fixed-base simulator, moving the position of the interior edge lines of a roadway altered the lane position of the vehicle while the same movement of the exterior edge line had no effect (Coutton-Jean et al., 2009).

The process of negotiating a curve begins well before the driver enters the curve. During the approach segment of the curve in the Shinar et al. (1977) study, the fixations of the eye were centered about 2 to 3 sec prior to the curve at the 60 mph (97 kph) driving speed tested. For the left-hand driver, eye lateral excursions were biased to the left curve while for the right-hand driver, eye excursions were biased to the right curve. Kelly et al. (2006) found biases in the direction of the curve of up to 5 deg. These data support the notion that oculomotor control is influenced by steering behavior in coordination with other non-visual inputs from head and neck sensors and driver trunk angle (Chattington et al., 2007). Steering, in brief, requires that the driver look toward the direction of intended vehicle movement.

The specific elements of the road used for steering guidance have been investigated extensively by Land and associates (Land and Lee, 1994; Land, 1998; Land and Horwood, 1996). Eye movement data occur within 1 deg of the curve tangent. Moreover, drivers normally explore upcoming road curvature well before they enter the curve, as noted earlier in this chapter. In the few seconds prior to entering the curve, perceptual judgments concerning curvature extent and preparation for steering and braking inputs need to be made. Driver judgments of curvature are, however, known to be poor. In a study by Fildes and Triggs (1985), magnitude estimates of the length of the arc of a curved road were compared to a standard view. A reduction in actual curve radius, rather than decreasing driver perceived radius of the curve, actually increased perceived radius of the curve. A decrease in the angle of the curve led to what the authors describe as a *perceptual flattening* of the curve. Sharply curved bends with small curve angles were perceived as being less curved, while the reverse was true for less curved bends.

The perceptual illusion of underestimating the curvature of sharp bends impacts the judgment of both entry speed and steering angle necessary for negotiating the curve of sharp bends. Excessive entry speed is known to be causal for road departures in curves. The excessive speed makes it more difficult to correct the steering wheel angle anticipated for negotiating the curve.

Decreasing road radii results in the illusory underestimation of road curvature even though, physically, the road curvature is becoming more severe. The perception of bends has been classified by Milleville-Pennel et al. (2007) as severe, those roads with 100–160 m radii, and as very severe, those roads with 26–80 m radii. In their study, only the latter, or very severe roads, were underestimated in curvature. The very severe category is the most likely one to present steering difficulty due to excessive speed while approaching the curve.

BEARING ANGLE OF LEAD VEHICLE

Apart from the delineated road path and optic flow field, drivers may also use the vehicle ahead, if available, as a guide for steering control (Salvucci and Gray, 2004). To be of use to the driver on a steering control, the vehicle would need to be in the 2- to 3-sec anticipatory control window discussed earlier. In curve negotiations, steering control is dependent on the bearing of the vehicle ahead relative to the driver's own vehicle. This adds to the task loading as a safe vehicle following distance as well as the maintenance

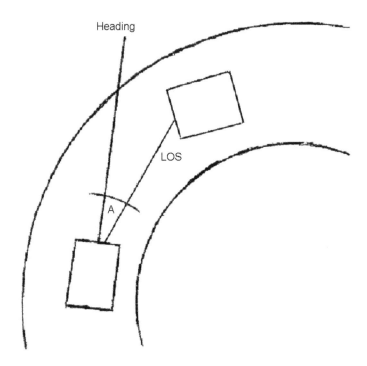

Heading

LOS

A

FIGURE 2.2   Lead vehicle bearing angle (A) and line of sight (LOS).

of lane position. This use of a lead vehicle may require a shorter headway than may otherwise be desired.

The angle formed between the lead vehicle and the following vehicle is shown in Figure 2.2. As with the use of the tangent point in curve negotiation, the bearing angle can be viewed as a part of an oculomotor response component of the driver, which includes both visual and non-visual components. The bearing angle needs to be continuously updated by the driver in order to extract accurate anticipatory information. This usually means that the driver needs to reduce the distance from the lead vehicle in order to extract useful anticipatory information. Note that the bearing angle of the lead vehicle may or may not be the same as the tangent of the curve depending on the position of the lead vehicle at any one time.

Evidence of short headways of a lead vehicle for steering guidance has been shown in a study of driving behavior in reduced visibility (e.g., fog) by Saffarian et al. (2012). Another study of driving behavior in fog showed that the absence of the lead vehicle leads to less accurate steering performance (Laurent and Gabbiani, 1998). Drivers in these studies reported that the short headway required for the use of bearing angle in guidance increased the feeling of risk of collision. Rear-end collisions in these circumstances become more likely when headways are reduced below 2 sec as the time needed to bring the driver's vehicle to a stop.

Nonetheless, the driver may feel the need to follow the lead vehicle more closely under reduced visibility in order to take advantage of the steering guidance provided. In this case, the driver has made a decision, conscious or otherwise, that the risk of rear-end collision is preferable to the loss of guidance information and road departure that may result. This may also account for the tendency of drivers to maintain higher-than-desirable speed in fog in order to maintain contact with a lead vehicle though the higher speed may contribute to the risk of both collision and loss of control (Yan et al., 2014). Risk tradeoffs in driver behavior will be discussed in more detail in Chapter 4.

SPLAY ANGLE AND SPLAY RATE

An additional perceptual cue that could support steering control is provided by the combination of the road centerline and road edge markings. When the vehicle's lane position is varied with steering or by forces such as cross winds on the vehicle, changes in the splay angles projected in each lane can be detected by the driver. The change in the relative size of the displacement of the projection provides the driver with the stimulus for steering control inputs. The speed at which the splay angle changes or *splay rate* determines the speed or steering response needed to make the lane position correction.

An experiment by Li and Chen (2010) attempted to identify the relative contribution of splay angle optic flow and bearing angle on steering control. Where bearing angle could be used in environments where there are no road markings and little optic flow, optic flow and splay angle information were the most valuable in steering control under straight road conditions. Furthermore, the authors suggest that the splay angle is generally more robust for lane position control tasks as unlike the bearing angle, it is unaffected by vehicle rotation. In the study by Chatziastros et al. (1999) in simulated tunnel driving, optic flow field information was the primary factor in maintaining lateral control while lane border information improved steering accuracy. As with other perceptual cues that aid steering control and other driving behaviors, splay angle and optic flow appear to be additive in their effects. Even with clear road markings, the splay angle changes can aid lane position control while enriching the optic flow field aids perceived lane position even beyond that provided by the splay angle.

STEERING CONTROL BEHAVIOR

The perceptual processing by the visual and non-visual systems of the driver is only part of the steering control system. The second component is the steering control input response required to adjust the vehicle heading when required. The steering wheel angle (SWA) input by the driver has an effect on the angle of the front tires, which is proportional and fixed. While the ratio of the SWA to the front tire angle varies from one vehicle to the next, the typical ratio for a passenger vehicle is 1:15. That is, for every 15 deg of SWA change, there is a change of 1 deg of tire angle. Drivers learn for each vehicle they drive the amount of SWA needed for a given rate of turn. The amount of force or torque applied to achieve a given SWA is also dependent on the vehicle although most passenger car steering torque is in the range 4 to 5 Nm. With power assist functions, the amount of steering torque required is substantially reduced from that required of older vehicles without power assist. The amount of power assist varies depending on the amount of torque input required. This typically means larger assists when more aggressive steering is required as in parking and smaller assists when it is not as at highway speeds.

A driver's steering control behavior at the level of physical force control of the steering wheel is driven by visual inputs from the road environment. The driver must extract this input in real time in order to initiate and update steering wheel inputs as needed. While the visual system of the driver including the oculomotor system that drives eye movements is essential, it also includes other driver body components including head, neck, and torso movements. Thus, steering is best characterized as a complex perceptual-motor chain of neurophysiologic and neuromuscular events which culminate in precise steering wheel movements. These movements are initiated so that sufficient force is applied to produce the desired effect but not so much as to result in oversteering.

The perceptual-motor chain involved in steering is one of a variety of perception–action sequences that drivers learn. In driving, the act of steering a vehicle is repeated over many hours of driving to the point where the perception–action sequence becomes automatized. Automatization was introduced in Chapter 1 as a general phenomenon of human behavior found in many diverse activities including driving. It is essential not only to produce precise and repeatable actions but also to minimize dependence on driver attention. The precise number of hours required of drivers to achieve automatization is unknown though it is likely to occur within the first few years of driving.

Studies of steering behavior suggest that the perception–action sequence becomes quite precise when drivers need to negotiate curves. Land (2006) describes basic steering behavior such as that described above as a gaze-action system, i.e., a perception–action sequence controlled by the driver gaze where the gaze movement leads or anticipates other body movements. For steering in curves, the gaze is turned toward the tangent point of the curve and remains in the same angular position through the curve. The path of the SWA and gaze angle is similar to the reciprocal of the radius of the curve (1/r). The SWA is directly related to the square of the gaze angle and inversely related to the distance to the road or lane edge.

In the motor control model proposed by Chattington et al. (2007), the gaze of a driver leads to an oculomotor perception–action sequence. Additional inputs from other neuromuscular systems, such as the neck and torso, are under higher speed curve driving. In this case, the neck provides the motor feedback with reference to the direction of the gaze while the torso provides motor feedback with reference to the vehicle's heading.

The emphasis of these models is on the compensatory tracking that drivers engage in curve driving. Additionally, the focus is on more serve curve radii (<100 m), which are more demanding of the compensatory component of driving. The anticipatory or lead component of steering in curves that occurs a few seconds prior to curve entry prepares the driver for the demands of compensatory tracking that follows.

The research of Frissen and Mars (2014) reveals that the anticipatory period prior to the entry into the curve is essential for steering accuracy. By systematically visually degrading road information required for antici-patory and compensatory processing, they were able to show that anticipa-tory information was more sensitive to visual degradation. This is likely due to the need to process higher image details of road curvature at greater distances. Without this information, the driver was left to react rather than anticipate. This, in turn, resulted in late entries into the curve and with the vehicle from the inner road edge or lane is desirable. Steering became more erratic and lane position more variable.

Similar anticipatory behavior of drivers found in curve driving occurs in the lane change maneuver as well. Experiments of Salvucci and Liu (2004) reveal that an anticipatory period prior to lane change is character-ized by drivers visually scanning the area for traffic conflicts and then shifting the visual scan to the destination lane. At lane crossing, the drivers exhibit a sine wave pattern of steering with a tendency to steer away from the destination lane 2 to 3 sec prior to lane crossing. However, a study by Doshi and Trevedi (2009) analyzed the driver's eye and head movement in lane change and found that the latter was a better prediction of the actual driver lane change decision in the 3 sec prior to executing the maneuver.

Steering control is a more complex process than might be expected from a seemingly simple perceptual-motor activity. The control process itself is expected to become automatized with driver experience. The perception–action sequence is preceded, however, by a period of anticipatory behav-ior initiated by selective attention to a particular task goal. In the case of the curve driving task goal, an initial assessment of road curvature occurs prior to curve entry with the compensatory tracking component following. If the driver correctly anticipates curvature, the compensatory tracking occurs smoothly, requiring little or no conscious attention. Similarly, lane changes are preceded by an anticipatory period in which the driver attends to the task goal of leaving the current lane and entering a new one. The lateral movement of the vehicle appears to be anticipated by head movement preceding a change in gaze direction by several seconds. Visual scanning for traffic conflicts then follows with the actual execution of the maneuver proceeding in a compensatory fashion comparable to curve driving.

SUMMARY

Essential to safe driving is the control of vehicle speed and guidance. The visual perception of vehicle speed is determined largely by the optic (retinal) flow of road elements that occurs as part of the optic array when the vehicle (driver) is in motion. Optic flow is affected by the density of the perceptible elements of the roadway, by driver eye height above the roadway, and by the field of view available through the vehicle structures. The edge rate of road elements across the fixed points of the windshield affects the perception of speed. Road luminance and contrast level strongly affect the perception of vehicle speed. Non-visual elements also affect the perception of vehicle speed including wind and engine noise as well as vibrations transmitted by tire travel over road surfaces. Guidance or steering control in straight road segments depends on ambient, pre-attentive visual perception of optic flow patterns in the visual periphery, which allow for positive lane position control. For curves, guidance depends on driver tracking of the tangent point of the curve, which changes dynamically as the curve is negotiated. The ambient, pre-attentive control of lane position and the automatized, closed-loop, perceptual-motor control of steering are both essential to the management of the driver's attentional resources in driving.

3 Perception of Distance, Time-to-Collision, and Collision Avoidance

In the previous chapter, factors affecting the control of vehicle speed and guidance under a variety of road conditions were discussed. Identifying and utilizing perceptual cues, primarily visual, are tasks that must be mastered by all vehicle operators. Both tasks are dependent largely on the extraction of information from the optic array, such as optic flow and edge rate, and on visual cues like the tangent point of a curve that provides steering information.

The identification and avoidance of a collision hazard is the third critical task that drivers must master. While control of vehicle speed and guidance are essential to achieve the goal of traveling from one point to another, avoiding a collision is a necessary component of vehicle travel. The collision hazard is defined not simply as the avoidance of colliding with another object on the road, such as another vehicle or a pedestrian, but also as the avoidance of a situation or event which may result in a collision. These might involve the driver entering an intersection against the traffic signal when cross traffic is present. This situation exposes the driver to a collision with cross traffic entering the intersection. Departure from the road surface when objects (trees, structures, etc.) are present is also a situation where a collision is likely to occur. The important element in collision avoidance behavior is that it is a response to an *imminent* hazard. An imminent hazard is a hazard that will occur if the driver does not make an avoidance response (e.g., braking or maneuvering). This definition is in contrast to a *potential* hazard which may or may not result in a collision or loss of control sometime in the future. A potential hazard does not require an immediate driver control response but does require that the driver monitor the hazard as it may at any time evolve from a potential hazard to an imminent one. Potential hazard recognition and related behavior will be addressed in the next chapter (Chapter 4).

STATIC VISUAL CUES TO DISTANCE

The identification of and response to an imminent collision threat requires both distance judgment and the ability to determine whether an object or situation represents an imminent hazard. The perceptual cues supporting this task are heavily dependent on the fundamentals of human vision in depth and distance perception. While a more detailed discussion of distance

DOI: 10.1201/9781003454373-3

perception was provided in Chapter 1, a brief review of visual cues supporting distance perception in the driving task domain will be provided here.

Pictorial cues to distance for the driver are those that are present in the road
environment at any one time. Not all cues are present at one time, nor are the
cues present at any one time necessarily the most reliable in the perception
of distance. This means that a driver's ability to determine the distance of a
hazard will be determined by what specific cues are available at that time. As
the road environment necessarily limits the number of distance cues available,
only those will be discussed in this chapter. Second, the concern here is for
those distance cues that are relevant to the issue of collision avoidance, so the
discussion is restricted to those cues involved at relatively short ranges. These
ranges are determined by the time needed by the driver to recognize the hazard and to respond with either braking or maneuvering. Unlike the potential
hazards to be discussed in the next chapter, avoidance of an imminent collision
hazard is characterized as a *reactive* response limited to relatively low-level
perceptual-motor sequences. This means that higher-level cognitive monitoring and decision-making processes that require additional processing time of
the driver are not involved or are minimized.

TEXTURE-DENSITY

Among those cues most likely to be involved in distance estimation for
collision avoidance is texture-density cue. This visual cue to distance
occurs when a surface, such as the surface of the road or highway, presents
a regularly recurring pattern of elements that increase in density as the
distance from the driver increases. While the texture-density cues that support driver distance perception occur primarily in hard-surfaced roads of
asphalt or concrete, off-road vehicles traveling grasslands also need these
cues. This is due to the strong texture-density cues that grasses provide as
normal patterns of growth.

The strength of texture-density cues for roads and highways is dependent on the specific components of the surface material used in their construction. Typically, aggregates such as granite are mixed in with asphalt
or concrete as a strengthening agent. The individual aggregate elements
will vary in size, reflectivity, and density.

The aggregate used in roads needs to be of sufficient size to allow the
driver with typical visual acuity (20/40 or 6/12 m) to view it at a distance
adequate to avoid collision. A large aggregate of 1.9 cm (0.75 in) width
would need to be within 33.2 m (109 ft) with suprathreshold contrast to be
detectable by a driver. This would only be the case if the object were seen
at driver eye height with the driver's line of sight parallel to the surface.
However, the driver is viewing the road surface object at an angle relative to the surface with the object itself embedded in the topping material.

This will result in an aggregate that subtends less visual angle and therefore a lowered likelihood of visual detection and processing. It would be expected that the effects of texture-density on distance perception are probably somewhat less than 30 m (100 ft).

The contrast and reflectivity of the aggregate relative to its immediate background is necessary for driver visual detection and processing of texture-density cues. Reflectivity is particularly important if the aggregate is to be seen under mesopic driving conditions. Reflectance from headlights, which typically extends some 150 ft at a low beam setting, will make the aggregate stand out in comparison to its topping.

Alternative means of improving road texture-density and therefore driver distance cues are available. Some of these were explored in the previous chapter (Chapter 2) with respect to the control of speed and steering. Texturing road surfaces is known to improve accuracy in the perception of vehicle speed primarily by improving optic flow. However, the exact relationship between texture-density and distance perception on roads is not known though mathematical tools that can quantify texture-density are available (Srinivasan and Shobha, 2008). It is that the empirical research necessary to define the relationship between various levels of texture-density of the road and driver distance perception has not yet been conducted. Thus, while it is likely that some texturing of roads will influence distance perception in collision avoidance, the optimal level is not known.

SIZE–DISTANCE INVARIANCE

The relationship between the visual angle subtended by an object to the perceived distance of that object by an observer is the essence of the size–distance invariance in distance perception. This is arguably among the most powerful distance cues as it expresses a fixed relationship between an object's retinal size and its distance. The larger the retinal size of the object, the closer the object is to the driver. Objects that are moving away from the observer will have retinal sizes that grow smaller. With experience, observers develop from this retinal image an estimate of the distance of common, familiar objects they repeatedly encounter. This applies to drivers as well, particularly with regard to vehicles and pedestrians. The calculation of the visual angle is as follows:

$$V = 2 * \arctan (S/2D)$$

where
   V = visual angle (degrees) subtended
   S = the longest side of the object
   D = distance of the object from the driver

It was noted earlier that observers tend to underestimate distances when asked to make absolute metric judgments. In optimal conditions with strong texture fields, distance judgments by observers beyond about 66 ft (20 m) begin to increasingly underestimate actual distances (Norman et al., 2020). This is likely due to the optical phenomenon of visual compression, which occurs with increasing intensity as distances increase. While equivalent measurements of drivers in field conditions are unavailable, it is likely that drivers also underestimate distances beyond 20 m as the texturing of road surfaces is likely to be less than optimal.

HEIGHT IN THE VISUAL FIELD

Another visual cue to distance which is particularly important at the greater distances involved in driving is the relative height of an object in the visual field. As objects increase in distance for the driver, they also increase in height in the visual field. This is an optical phenomenon that is a result of increasing angular declination of the line of sight to the horizontal (Ooi et al., 2001). As seen in Figure 3.1, the angular declination of the line of sight to the eye level to the horizon increases as a distant object comes closer. This is a useful cue provided the driver's vehicle and object of interest are on a level road but less useful in the undulating crest and sag curves.

RELATIVE SIZE

The relative size of an object that is seen with other objects of the same class affects its perceived distance from the driver. When the objects of a similar size are seen together, the object that looks larger will be perceived as closer to the driver. Relative size cue is related to the concept of the *perceptual set* where exemplars of the set such as pedestrians are conceived as having comparable dimensions such as adult height. At a distance, a

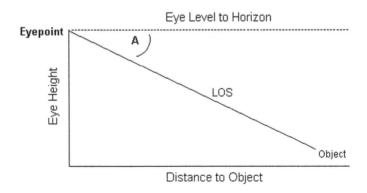

FIGURE 3.1   Angle of declination (A) and line of sight (LOS).

member of the set such as a child entering a street may be seen as an adult at a distance and interpreted as being at a greater distance from the driver.

OBJECT OVERLAP

An object that overlaps other objects in the visual field will always be perceived as closer. Overlap cues to distance take precedence over other cues such as relative size or size–distance invariance.

ATMOSPHERIC CONTRAST

Atmospheric contrast, sometimes termed aerial perspective, results from the diffusion of light reflected from an object as a result of airborne particulates. The effect of this diffusion increases with distance so that the contrast of an object to its background is also reduced. Atmospheric contrast is likely to be present at longer distances or under conditions where suspended particulates are particularly dense as in the case of pollutants such as smog. The effect is to make the object appear further away than it would otherwise be under clear air conditions. Contrast effects are exponential, not linear, with distance.

LINEAR PERSPECTIVE

A cue to distance that occurs in straight roadways due to their parallel lines is linear perspective. The relative position an object takes on the roadway which presents a linear perspective cue determines its distance in relation to other objects regardless of the retinal size of those objects.

MULTIVARIATE CHARACTER OF DISTANCE CUES

At any given moment in any given driving scenario, drivers may be exposed to a variety of distance cues. While some cues will provide unique distance information, other cues will present distance information that is essentially redundant with other cues. The redundancy provides the driver the advantage of not having to sample the entire visual field to provide needed distance information with respect to specific objects. For example, an oncoming vehicle will provide distance information in the form of retinal size but embedded in a roadway which provides both texture and linear perspective. In perception, cues to distance are additive so multiple cues tend to increase the accuracy of distance judgments.

However, it is also the case that some visual cues to distance have an inherently higher correlation with distance than others. This higher *ecological validity* of a cue is likely developed with driver experience but may well exist as a product of experience in the non-driving environment as well. It is likely that size–distance invariance is one such visual cue of distance. Drivers may weigh visual cues to distance based upon the driver's experience in how well these individual cues accurately predict the distance of a collision hazard, for example, and rely on these cues more heavily than others. In this way, a driver can be more efficient in the employment of resources such as visual attention in collision avoidance.

VISUAL MOTION CUES TO DISTANCE

Thus, far only static cues to distance have been discussed. Static cues, sometimes called pictorial cues, avail themselves when the observer and cue are in a fixed position. In driving, however, it is common that either or both the vehicle and the object being viewed are in motion. When the observer or object is in motion, the perceived distance depends on visual cues provided by the optic flow components available. These are non-pictorial cues as they are effective as cues to distance only when either the observer or the objects are in motion.

SIZE–DISTANCE RELATIONSHIP

An object that increases in perceived size from one visual sample to the next will be perceived as moving closer even if the actual movement of the object is not seen. For drivers viewing familiar objects in the roadway, prolonged viewing may not be possible or desirable. In this case, periodic visual sampling of the object may be a more efficient use of vision resources, particularly for more distant objects such as approaching vehicles. This is simply a version of the size–distance invariance pictorial cue applied to objects in transit from one position to another.

BINOCULAR DISPARITY

The disparity produced by the slightly different images projected to each eye can be used to detect oncoming objects. This occurs due to changes in disparity that occur with distance from the object. Binocular disparity as a dynamic cue to distance, however, appears to be limited to close objects. Distances that might be affected by disparity cues appear to be on the order of 6 to 10 ft (1.8 to 3 m) (Schiff, 1980). Some evidence for this limited use of binocular disparity is found in Gray and Reagan (1998). In this study, the accuracy of time-to-collision (TTC) estimates improved with the availability of binocular disparity but only at a short range of 5.2 ft (1.6 m). The limits of binocular disparity would likely be a useful cue to drivers only in maneuvers requiring the avoidance of objects very close to the vehicle such as those in parking. At higher road speeds, some closure rate information may occur at the last moments from a collision with another vehicle or pedestrian, but the short distances involved may render the information moot as the time for the driver to react would be too brief.

CONVERGENCE AND ACCOMMODATION

As objects become close to an observer, the eyes will automatically converge on the object being tracked. Accommodation or changes in lens shape will occur during this time in order to eliminate blurring of the oncoming age. These cues are useful only at very short distances (e.g., less than 2 ft or 0.6 m) and therefore are likely to be useful as distance cues only within the vehicle.

MOTION PARALLAX

Another visual phenomenon that occurs within an optic flow can provide information on the distance of moving objects and their relative motion to the driver. Motion parallax is the translational movement of scene elements in relation to the direction of the driver and therefore of the driver's vehicle. Closer scene elements will move faster in relation to scene elements than those further away. The most dramatic example of motion parallax can be seen by a driver viewing the movement of the scene as it passes a side window. Areas above the point of fixation will be seen as moving in the same direction as the driver's vehicle, while the scene movement below the point of fixation will be seen as moving in the opposite direction. Motion parallax and binocular disparity effects on driving behavior are important issues in the design of driving simulators, which do not normally provide either of these cues to distance (Lee, 2017). Attempts to simulate head movement parallax in the overtaking maneuver have not thus far shown that head movement parallax affects the overtaking maneuver (Eriksson et al., 2015).

TEXTURE-DENSITY AND OPTIC FLOW

Texturing of the visual field that appears in the foreground of an object such as a vehicle or pedestrian is a known static cue to distance. Higher texture-density will make an object appear closer even though neither the object nor the observer has physically moved. During movement of the driver, the optic flow may be enhanced by increases in texture-density, particularly on the road surface. This increase of optic flow intensity with an increase in texture may affect perceived self-motion and increase the accuracy of detecting changes in lead vehicle distance in following tasks. As noted earlier in Chapter 2, increases in texture-density impact self-motion perception due to the increase in optic flow that it creates. However, its role as a motion-in-depth cue for collision avoidance is less clear. One such possibility occurs in the detection of changes in the speed of the lead vehicle while following.

OBJECT EXPANSION AND CONTRACTION

The optical expansion or contraction of an object as it moves closer or away from an observer is a phenomenon inherent in the retinal flow of objects that occurs during movement. Expansion and contraction of the retinal size of objects inform the driver of their dynamic relation both with the driver and relative to each other. The most important visual motion cue for the detection of collision is the expansion of the retinal size of an object as the distance between it and the driver is reduced. For collision avoidance, a number of attempts have been made to relate this phenomenon to the problem of driver estimates of the TTC with roadway objects.

TIME AND DISTANCE ESTIMATION IN COLLISION AVOIDANCE

How well drivers estimate distance in a driving task is a question different from that of driver estimates of absolute metric judgments of their distance from objects in the road environment. Nominally, judgments where drivers have to provide a numerical value for a perceived distance to the object require a skill that drivers normally gain neither by experience nor by training. It is not surprising therefore that those drivers are unable to determine the absolute distance of objects either on or near the roadway (Burney, 1977). In general, the results from this study reveal the tendency for drivers to underestimate absolute distances. Drivers, moreover, tend to be more accurate at relative judgments, and those judgments increase in accuracy at shorter distances. Thus, judging the relative distances of two vehicles approaching in a multilane roadway enables the driver to decide on which vehicle is likely to present the shorter TTC. However, the estimated TTC of either will still be inaccurate. Underestimating TTC is common for drivers under a variety of conditions. The size of this under-estimation of TTC varies widely (2% to 40%) depending on the particular method that is used to assess it (Gray and Reagan, 1997)

ESTIMATING TIME-TO-COLLISION

However, as will be seen, it is not absolute distance as such, but the time available to execute a maneuver that is the most important factor that a driver must consider in collision avoidance. Drivers executing maneuvers such as overtaking traffic, turning in front of approaching traffic, approaching intersections, and following other vehicles need to estimate the time-to-arrival (TTA) or time-to-collision (TTC) of an oncoming vehicle in order to determine what action is needed and when. This time is added to the time to execute the maneuver plus some safety factors to determine whether the maneuver is safe and effective to execute. Inaccuracies in estimating time in any of these factors can lead to a collision.

Among the most studied elements of driver behavior has been their ability to judge TTC with other objects on or near the roadway. In this section, TTC estimates (TTCest) will be examined as a function of different task scenarios as each such scenario will differ in the kind of information and the prevailing conditions presented to the driver. These scenarios include those where either or both the driver's vehicle and the target vehicle or object are moving. Variations of this include vehicle-following tasks where the driver attempts to avoid a rear-end collision when the lead vehicle brakes. Another is to avoid collision when a vehicle or object is stationary in the roadway.

Objectively, TTC is calculated as the *closure rate* or rate at which the distance between two objects is closed or eliminated in the course of time. The closure rate is simply the distance divided by the elapsed time given a constant velocity of the closing object if the target is stationary. If both objects are moving toward each other, then the closure rate is the sum of each object's velocity. In the case of two vehicles approaching one another, as might be the case of an overtaking maneuver into an opposing lane, the closure rate is therefore halved when compared to the approach to a stationary object. For example, two vehicles approaching each other, each traveling at 60 mph, will have a closure rate of 176 ft/sec. In this case, the time available to avoid a collision is reduced by half when compared to the approach to a stationary object.

As the actual closure rate is not normally available to the driver, an estimate of TTC needs to be derived by other means. Perceptually, this means that the driver must estimate TTC using whatever visual information is available at the time. Whether an object is moving toward the driver or the converse, the distance between the two may be estimated either by extracting the visual cues presented from an object continually changing in perceived size or by other visual distance cues. In the case of approaching objects, periodic sampling of an object's distance can provide momentary estimates of TTC to the driver. Thus, a strategy of quickly eliminating a collision hazard based upon a momentary estimate of TTC using only retinal size, for example, can provide a means to minimize the expenditure of resources on traffic which does not present a collision hazard. As discussed earlier, observers use their familiarity with object sizes at various distances to estimate how far a given object is from them.

The ability of drivers to estimate TTC has been investigated both in driving simulators and on roadways for decades. A variety of variables have been evaluated for their impact on TTC$_{est}$ including target vehicle speed and size, speed of the driver's vehicle, road lighting conditions, the particular task or maneuver (e.g., following, passing), and driver task loading, among others.

TIME-TO-COLLISION IN DRIVING TASKS

As it is easier to understand driver behavior within the context of a given task, driver estimates of TTC will be given by driver task. A discussion of the task structure of each of these driver tasks or maneuvers will be described, followed by a more detailed analysis of the perceptual and cognitive constituents believed to support the collision avoidance task.

Turning in Front of Oncoming Traffic

Turning across the path of oncoming traffic at an unprotected (unsignaled) intersection is one of the more hazardous tasks drivers must perform. The turn can be executed from a full stop or while the driver's vehicle is moving. In either case, the driver needs to determine the time available to execute the maneuver with the addition of some safety margin. The key to the turn maneuver is largely dependent on the driver's ability to estimate the arrival time of oncoming traffic at the intersection. This time is then added to the estimated time to complete the road crossing.

The studies of TTC$_{est}$ of oncoming traffic in the turn maneuver reveal a number of variables affecting the estimate. In reviewing prior studies of TTC$_{est}$, Caird and Hancock (1994) noted that underestimates of TTC tend to increase as TTC itself increases. In other words, TTC$_{est}$ becomes more accurate when the oncoming vehicle is closer to the driver. Moreover, tangential paths of an oncoming vehicle were more accurate than head-on approaches. Tangential paths are a characteristic of vehicles approaching in the lane opposite the driver and are likely to yield more accurate TTC$_{est}$ than those of oncoming traffic as would occur in the overtaking maneuver (see below). A reduced visual field such as monocular vision reduced the accuracy of TTC$_{est}$ while the increased velocity of the approaching vehicle increased accuracy. In the authors' study, estimates of TTC in a left turn were strongly influenced by vehicle size. This was particularly true when comparing estimates of motorcycles to those of cars and trucks. Smaller vehicles such as motorcycles resulted in less underestimation of TTC than larger vehicles. Overall, as found in other studies, drivers consistently underestimate TTC and the degree of underestimation increases with the TTC. The only exception to this was for vehicles with a TTC under 1 sec.

In this case, drivers *overestimated* TTC. Fortunately, the consistent under-estimation of TTC by drivers means that the drivers are judging the arrival of the oncoming vehicle into the intersection sooner than the vehicle actually does. This, in effect, adds an in-built safety measure for collision avoidance in addition to others that may be added by the driver.

The reduced underestimates of TTC by drivers when smaller vehicles, such as motorcycles, approached were replicated in a more recent study by Horswill et al. (2005). The authors also tested the hypothesis that small vehicles are judged to arrive sooner than large vehicles due to the fact that the looming threshold had not been achieved when the TTC estimate was made. The looming threshold is the point at which the vehicle's retinal size begins to expand as it approaches (see below). However, the differences in TTC_{est} between cars and motorcycles continued even when the latter had exceeded the looming threshold.

Vehicle Following

Unlike the turn maneuver across the path of oncoming traffic, following traffic ahead presents the driver with the problem of continuously maintaining a safe distance from the lead vehicle when both vehicles are in constant motion. The lead vehicle may increase or decrease speed at any time. If the driver wishes to maintain a constant distance, speed needs to be altered to match the lead vehicle's speed. Vehicle following is a routine task in driving but presents a collision hazard if the following distance does not allow adequate time to stop should the lead vehicle decelerate quickly. While drivers have been taught to maintain safe separation times (e.g., 2 or 3 sec), studies of actual following distance maintenance indicate that drivers often maintain shorter following distances than is safe. In a controlled track study of following distances in the U.K., drivers maintained a following distance of 1.9 sec (Colburn et al., 1978). They did so regardless of the speed of the lead vehicle, level of experience, or whether or not they were informed of the likelihood of the lead vehicle stopping. These results confirm those minimum safe distances from an earlier study by Rockwell (1972). A field study by Taieb-Maimon and Shinar (2001) in Israel found the following times to average only 0.66 sec (SD = 0.26) when drivers were asked to maintain a safe distance. When asked to maintain a "comfortable" distance from the lead vehicle, distances increased to an average of 0.98 sec (SD = 0.36 sec). The lead vehicle speeds ranged from 30 to 62 mph (50 to 100 kph). Fully 93% of the following times under the minimum safe distance instruction were less than 1 sec and none were more than 1.4 sec. For comfortable instruction, 98% of distances were less than 1.68 sec. Following times were constant across all speeds for both instruction conditions. Notably, driving manuals for these studies at the time they were conducted called for drivers to maintain a 2-sec following distance from the lead vehicle.[1] It is evident that the task of the vehicle following under a variety of speeds can be executed with consistency by drivers across a variety of speeds. However, the following distance that a driver uses will be affected by a variety of factors including training, traffic density, risk perception, and the extent of local enforcement of unsafe following distances. Perceptually, the following distance, as with other distance judgments, is routinely underestimated by drivers. However, the threat of a collision in following is usually a result of the unexpected deceleration of the lead vehicle. Detection of this deceleration and the rate of that deceleration by the lead vehicle are essential if the driver is to prevent a collision even if the following distance is within safe limits.

The detection of the deceleration of a lead vehicle is not problematic if the driver of the following vehicle is attentive. A study of eye movement behavior in a following task found that the lead vehicle was being tracked by about 79% of all eye movements recorded (Crundall et al., 2004). Moreover, only brief (1 sec) samples of the lead vehicle have been shown to be necessary to detect changes in speed (Sidaway et al., 1996). Manipulation of the size of the lead vehicle has been shown to result in reduced braking response time but it also reduced the following distance to the lead vehicle (Terry et al., 2008; Abdullah et al., 2013).

Among the more important perceptual variables that affect the following behavior of drivers is the angular subtense of the lead vehicle on the following vehicle driver's retina (see *Braking* section). This optical cue to distance as well as the change in the size of this cue is used by the driver to estimate the following distance as well the changes that may occur as the lead vehicle accelerates or decelerates (Andersen and Sauer, 2007).

Not surprisingly, the ability to follow a lead vehicle safely will be affected by prevailing weather and lighting conditions. Fog is known to increase the likelihood of collision with lead vehicles in the following task. In a study of the effects of fog on the following task, both the ability to maintain a safe distance from the lead vehicle and the ability to respond to changes in lead vehicle speed are affected (Kang et al., 2008). High-density fog has the effect of reducing the contrast between the lead vehicle and its immediate surroundings. The effect of this lost contrast reduces the ability of lead drivers to detect distance cues and changes in the optical expansion of the lead vehicle that occurs when it decelerates.

Following behavior is not likely to be affected solely by visual cues but by other factors as well. For example, drivers may set the following distance from heavy trucks based on whether they can see the truck's rearview mirrors. This is a common instruction in driving schools based on the assumption that if the truck mirrors are visible the truck driver will be able to see the vehicle behind them. Furthermore, a review of heavy

truck following behavior in the U.K. motorways did not find "close" following distance (less than 2 sec) to increase for heavy truck following
distance. This should occur if angular subtense is the primary cue as
it is greater for heavy trucks than for passenger cars at the same distance (Yousif and Al-Obedi, 2011). Another field study by Sarvi (2011)
found that drivers of passenger cars will follow heavy trucks at greater
distances than passenger cars. In contrast, drivers of heavy trucks typically allow a greater following distance when following other vehicles.
The larger following distance adopted by heavy trucks corresponds to
the greater braking distance needed for heavy vehicles when compared
to passenger vehicles. A variety of factors may account for the greater
following distance of passenger cars when following heavy trucks. The
issue of appearance by following vehicles in the rearview mirror of heavy
trucks is one. There is also the possible perception of greater harm that
might result from a passenger car colliding with a truck when compared
to another passenger car.

OVERTAKING

Overtaking or passing a lead vehicle incorporates some of the main task
components involved in vehicle following as well as those involved in
estimating the arrival of an approaching vehicle in the turn across traffic maneuver. Overtaking can occur either with or without opposing traffic depending on the availability of multilane roadways. Two-lane roads,
where one lane is used for approaching traffic, is a road environment common in rural environments and in developing countries where the lack of
industrialization, vehicle ownership, and the cost of multilane road systems make these alternative types of roadways less attractive.

There are two critical visual perception elements for the driver in the
two-lane, two-way overtaking maneuver. The passing sight distance (PSD)
is the distance along the roadway that allows for the safe passage of a
lead vehicle. Roadway design uses this sight distance calculation to apply
restrictions to the use of road sections for passing. The use of signage or
road markings is then applied to inform the driver that passing in this road
section is unsafe. The most common use is applied to the restriction of
passing on hills and curves where oncoming traffic may be obscured by
terrain. Assuming that the PSD is sufficient for passing and there are no
restrictions, the driver then must determine the presence and collision risk
of any oncoming vehicle.

The second critical element for the driver is to maintain a safe following
distance from the lead vehicle while ascertaining whether and when it is
safe to pass. This may require the driver to move partially into the opposing lane in order to obtain a clear view of oncoming traffic. Prior to any

lane change, the driver will need to determine whether any following traffic is attempting to pass to avoid a collision from the rear.

The analysis of driving behavior in the overtaking maneuver is complicated by the employment of different driver control strategies. In a field study of overtaking behavior by Wilson and Best (1982), a variety of strategies were employed by drivers to overtake vehicles in their path. Some drivers would overtake an impeding vehicle without a pause while "accelerative overtakers" paused briefly then accelerated passed the impeding vehicle. A platoon of two or more overtakers termed "piggy-backers" will pass the impeding vehicle as a group. In this case, only the lead vehicle of the platoon would have the necessary visual cues available to judge the distance and approach speed of the coming vehicle or the following distance from the impeding vehicle; the remaining vehicles in the platoon are now dependent on the judgment of the platoon leader. The studies of overtaking have necessarily concentrated on the common strategy of following the impeding vehicle briefly then accelerating passed the impeding when the approaching vehicle was at a safe distance.

A simulator study by Figueira and Larocca (2020) provides some information regarding the approach and following distances on the overtaking maneuver. A key element in this study was the speed of the lead vehicle which varied along with the approaching vehicle distance. The lead vehicle speed at the beginning of the overtaking maneuver had a greater effect than either its size or the approaching vehicle distance on the behavior of the driver following. At a speed of 37 mph (60 kph), the distance of the following vehicle from the lead vehicle ranged from 232 ft (70.8 m) to 266 ft (81.3 m) or a following time of between 4.3 and 4.9 sec. The distance of the approaching vehicle at the onset of the maneuver ranged from 1463 ft (446 m) to 1837 (560 m). At a closing rate of 108 ft/sec between the overtaking and approaching vehicle, the driver would be allowed between 13.7 and 17 sec to complete the maneuver. The net effect of increasing the PSD was to reduce the following time.

Both speed and overtaking strategy determine the following distance of the vehicle in the overtaking maneuver (Crawford, 1963; Wilson and Best, 1982). Isolation of the perceptual processes involved in overtaking requires that these factors be controlled. In a driving simulator study, drivers were found to initiate an overtaking maneuver when they determined that the oncoming vehicle was beyond a certain critical distance even though that distance was often insufficient for a safe overtaking maneuver (Gray and Reagan, 2005). These authors also found that following time was also affected by speed adaptation resulting from prolonged driving on a straight road. This adaptation added to the unsafe following distances maintained by drivers from the lead vehicle.

Braking

The rate of change in retinal size is a known visual cue that can be used for TTC_{est}. As an object approaches the driver or the driver's vehicle approaches an object, the retinal size of the image increases as the rate of closure increases. This is considered a purely optical phenomenon which does not require the driver to impose any higher-order cognitive functions to derive TTC_{est}. Among the first to quantify this phenomenon in the braking maneuver was Lee (1976). Drivers in this model, the *tau hypothesis*, use the ratio of two components in calculating when and how much braking should be used when approaching the rear end of a vehicle ahead. The first component is the angular subtense of the vehicle at a given point in time. This angular subtense (the visual angle subtended on the driver's retina) gives the driver the distance information to the vehicle being approached at a point in time. The second component is the rate of change in optical size over time or the rate of expansion per unit time. Thus, the TTC_{est} is an optical invariant (*tau*) expressed as the angular subtense of the object divided by the rate of expansion of the object over time. Tau is expressed in the following:

$$Tau = \theta / (\Delta\theta / \Delta t)$$

where:
θ = the angular subtense of the object on the retina
t = time

As tau is an optical variable, like optic flow, it exists as a pre-existing part of the visual perceptual system. As such, tau requires only the registration of the target object on the retina. Thus, the target must be within the area of the retina sufficient to detect a change in the angular subtense of the target. The denominator of the equation is the rate of change in that angular subtense over a unit of time (typically a sec). This threshold in the perceived onset of expansion or *looming* occurs when the rate exceeds 0.003 radians/sec or 10.6 arcmin/sec (Hoffman and Mortimer, 1996). The central area of the retina is sufficient for this purpose. Areas of the visual periphery may lack sufficient resolution to detect target rates of expansion at this level. In order to get an accurate tau, the observer needs to sample the oncoming object in the central visual field where sufficient resolution exists to perceive object expansion. This sample's estimate of TTC is accurate only if the approaching vehicle maintains a constant velocity.

A necessary ingredient of tau is the rate at which object expansion occurs, and the rate of expansion in retinal size will increase as the speed of closure increases. As tau is expressed as the inverse of the rate of change in object size, increasing the rate of change will reduce the size of tau and thus the estimated TTC. If tau is the visual cue needed to estimate TTC, then any increase in the rate of change or expansion will mean the observer will reduce the estimate of TTC. In other words, the estimate of a collision will occur with the target sooner due to the higher closing speed.

In a strict interpretation of Lee's (1976) model, the driver's entire estimate of TTC is in this single optic invariant, tau. Some important restrictions are notable, however. First, object retinal size has to be above the threshold of visual acuity to be detected as does the contrast of the object with its background. These variables have been discussed in earlier chapters. Second, the closure rate of the target must be a constant velocity. Changes in closure rate would invalidate any TTC_{est} from tau at any point prior to the change. This would require the driver to visually sample again to derive a new estimate. Third, the rate of retinal expansion must be above the looming threshold.

A key principal argument for the universal application of the tau hypothesis is that estimates are independent of the size of the target vehicle. As the formula for tau states, the TTC_{est} is the inverse of the ratio between the subtended visual angle and the change in visual angle at any moment. Thus, the physical size of the vehicle as such should not matter. Tau as an estimate of TTC should be the same whether the vehicle is large or small. If it can be shown that the vehicle size influences $TTC_{est,}$ then the tau hypothesis is false.

A number of studies of driver estimates of TTC have been conducted in order to test the tau hypothesis. In two studies that varied the size of the oncoming vehicles in several maneuvers, drivers estimated that smaller vehicles (e.g., motorcycles) would take longer to arrive than larger ones (e.g., trucks) in a turn across traffic maneuver (Caird and Hancock, 1994) and in overtaking (Horswill et al., 2005; Levulis et al., 2015; Levulis et al., 2018). This behavior is interpreted as the size arrival effect (SAE) and is dependent solely upon the rate of expansion of visual angle, not tau. As the rate of expansion in visual angle depends on the size of the vehicle, smaller vehicles have rates of expansion less than larger vehicles and will be perceived as arriving later than larger objects even though they are physically at the same distance and have the same approach velocity. Notably, the SAE of smaller vehicles does not appear to be due to a conservative response bias toward larger vehicles due to the harm large vehicles might cause in a collision (Levulis et al., 2018). The effect appears to be largely, if not exclusively, due to the differences in how the retinal size changes over time for smaller objects when compared to larger ones. Andersen and Sauer (2007) have provided evidence to support a driver model for vehicle following that relies solely on the visual angle and the rate of change.

It has been noted by some authors (DeLucia and Tharanthan, 2009) that tau and its derivatives including retinal size and rate of change in retinal size may prove difficult to apply to situations where a vehicle is repeatedly accelerating or decelerating as often occurs in the following task. Tau, for example, is only accurate when the lead vehicle followed is at a constant velocity so the following vehicle driver would need to wait until the lead had achieved a stable acceleration or deceleration speed before judging TTC. However, it would seem more likely that the driver would simply react to the looming cue during deceleration (absent a brake light cue) to avoid an imminent collision or to brake in response to the lead vehicle's onset of brake lights as a matter of course. The typical short distances of 1 or 2 sec in following would yield a strong suprathreshold looming cue during deceleration in either case. The DeLucia and Tharanthan (2009) and Andersen and Sauer (2007) studies of following behavior suggest that drivers use retinal size and the change in retinal size cue to respond to the lead vehicle's change in speed during following.

LOOMING SENSORY THRESHOLD

Both tau and SAE are dependent on when the observer perceives the increase in retinal size over time as it approaches. Looming occurs and is perceived as an increasingly rapid rate of object expansion. This is simply a consequence of the fact that the rate of expansion of an object that approaches an observer is not linear but geometric. Thus, as the vehicle and target object close in distance, the target object expansion rate begins to increase and this looming may trigger a strong collision avoidance reaction in the driver. This transition from slow to increasingly rapid expansion of angular subtense is shown graphically in Figure 3.2. Depicted is the size of three vehicles likely to be encountered in a roadway: a small motorcycle with a width of 2 ft (0.61 m), a full-sized passenger car with a width of 6.5 ft (1.98 m), and a heavy truck with a width of 8.5 ft (2.6 m). The changes in the rate of expansion are shown for two highway speeds: 30 mph (48.3 kph) and 60 mph (96.6 kph). Looming is presented in the figure beginning at 6 sec from the collision with the lead vehicle being stopped in the roadway. Note that in the case of the small motorcycle, at 6 sec and 30 mph, the retinal size is at 0.011 radians (0.65 deg) and thus has not yet exceeded the looming threshold of 0.003 rad/sec (0.17 deg) and will not do so until the driver is within 3 sec of impact. However, at 60 mph, this vehicle does not exceed the looming threshold until less than 2 sec from collision. Using a braking deceleration of the average driver under optimal conditions (15 ft/sec² or 4.6 m/sec²), heavy braking might prevent a collision in the case of the slower speed when looming is detected, but not in the case of the higher speed. The higher speed would require 5.9 sec to stop at 60 mph

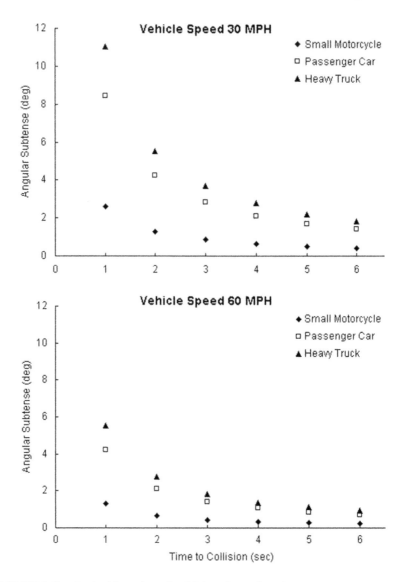

FIGURE 3.2   Optical looming of vehicles of varying size.

(88 ft/sec). In both cases, the object is of sufficient size to be detectable visually by a driver with a VA of 20/40 (6/12 m) under normal photopic conditions, but the driver will not experience the looming of such a small object at the higher speed until it is too late to brake. Only a last-second maneuvering would be possible to avoid a collision.

An additional element of this figure is the large differences in looming that occur with the larger vehicles when compared to the small motorcycle. For both the passenger car and heavy truck at 30 mph, looming increases at 8 to 10 times the rate of the small motorcycle in the 3 sec prior to the collision, with the greatest looming occurring in the last 2 sec. At 30 mph, the average driver would take 2.9 sec to brake to a complete stop. Strong looming begins for the larger vehicles at about the time needed to avoid a collision by heavy braking. Exceeding this threshold does not guarantee that braking will begin as other factors affecting the driver's response to looming may speed or delay action. Thus, using a looming threshold as a trigger for action, as is often used in forensic practice, is likely to be misleading. In fact, the looming threshold is often adjusted to accommodate accident data by factors of an order of magnitude or more. However, the looming threshold is a sensory threshold of the driver's visual system response to a purely optic phenomenon. The threshold itself is the midpoint of a distribution of driver thresholds that vary around the average. While this variation is likely to have an effect on an individual driver's specific threshold, the size difference would be in the range of fractional increases, not the orders of magnitude often cited in forensic reports. The looming threshold ought to be thought of as the beginning of the perceptual-action sequence in which drivers are accumulating evidence to support the response decision rather than as a purely reactive stimulus to a pre-programmed response.

Looming as a means of triggering action to avoid a collision requires that the driver interpret looming early enough and apply sufficient braking force to avoid a collision. Factors such as the size of the stopped vehicle and the relative speed of the driver's vehicle affect when looming begins. Since looming rates are greatest when the driver is nearest to collision, the driver needs to develop the ability to detect looming as soon after the looming threshold is exceeded.

ACCELERATED LOOMING THRESHOLD

While looming thresholds are known, it is less clear that the response threshold needs to be the same as the sensory threshold in the case of collision avoidance. Looming threshold can be and often is exceeded long before braking or maneuvering occurs. Indeed, the time necessary for the driver to respond would likely include, in some cases, not simply braking but maneuvering to avoid a collision. However, the forensic application of the looming threshold has found difficulties in fitting the looming sensory threshold (0.003 rad/sec) to field data regarding, for example, braking to avoid collision with a lead vehicle (Maddox and Kiefer, 2012). Instead, larger threshold values, an order of magnitude or more such as 0.02 rad/sec, are applied to fit emergency braking data.

The issue of such a large discrepancy between the looming sensory threshold and the actual time taken to apply full braking to avoid a collision suggests that a simple stimulus–response braking reaction is not occurring at the sensory threshold. Rather, in many cases only moderate levels or no braking at all will begin to occur when the looming sensory threshold has been surpassed. However, if the looming threshold does not yield a response by the driver to brake or to maneuver then the driver may be obliged to do so when some other more urgent collision stimulus is presented. Thus, the failure to respond to initial looming will often result in emergency braking or radical maneuvering only at the very last moment when the collision is imminent. For example, the stimulus to react would most likely occur when the looming of the lead vehicle has increased at a geometric rate. This likely occurs for passenger cars at a speed of 30 mph for a vehicle stopped in the roadway (Figure 3.2). The driver's retinal image size for the vehicle exceeds the looming sensory threshold 77.6 sec prior to collision while exceeding the accelerative looming value of 0.02 rad/sec 7.4 sec before the collision. Both cases allow adequate braking time to avoid a collision at this speed.

This explosive expansion in looming of a passenger car can be compared to looming of the small motorcycles at the same speed in this figure. In this case, the looming sensory threshold has been already passed at about 16 sec TTC, allowing more than enough time for the driver to respond. The accelerated looming threshold of 0.02 rad/sec begins at 2.3 sec TTC, with retinal size expansion doubling from 1.3 to 2.6 deg just one sec prior to impact. This comparatively late looming threshold coupled with the relatively low expansion rate of the retinal image for small vehicles as opposed to large ones may be one reason smaller vehicles like motorcycles are more likely to be involved in rear-end collisions than larger vehicles. The level of urgency elicited by the looming of the small motorcycle was simply not sufficient to elicit the necessary braking response in time to prevent a collision.

However, the vulnerability of smaller vehicles to rear-end collisions due to the lack of expansive looming does not explain the need for forensic manipulation of the looming sensory threshold to account for braking behavior in field data of rear-end collisions. Investigations of the looming of an object in the visual field have shown that it attracts attention in laboratory studies (Franconeri and Simons, 2003; Muhlenen and Lleras, 2007), so the reason of drivers' failure to respond may lie elsewhere. A more plausible explanation separates the sensory threshold, a purely visual phenomenon, from the response threshold to an impending, imminent collision such as that which occurs when approaching a stopped vehicle. The braking response to the looming threshold appears to be delayed even though the driver may sense that the retinal image of the stopped vehicle is beginning to expand. For these drivers, there is no sense of urgency as the image has not achieved what might be termed the *accelerated looming* stage to separate it from the basic looming sensory threshold detection stage. Accelerated looming sufficient to trigger a wholly reactive, urgent braking response appears to occur at the very late stages of looming and is dependent on vehicle size. Larger vehicles will reach the accelerative looming stage earlier than smaller vehicles, and the rate of looming in the accelerated state will be greater. In the case of very small vehicles, such as small motorcycles or bicycles, the accelerative looming stage will be very modest in expansion and very late in the looming process. In some cases, this lack of time and urgency will be too little and too late to avoid collision.

The question then remains as to when this triggering occurs during accelerated looming. Extensive analyses of braking data from 116 rear-end crashes and 241 near crashes suggest that this occurs when looming reaches 0.02 rad/sec (Markulla et al., 2016). In these cases, drivers would have been exposed to looming sensory thresholds long before the triggering event, the accelerated looming threshold, occurred. Attending to the looming sensory threshold by a driver would have allowed for an early and less urgent braking or maneuvering response to the threat. Instead, the driver continued the approach until the threat became imminent and overwhelming. While the authors argue for an accumulative mechanism of looming that builds up to a triggering event, it is argued here that the result is that these drivers due to inattentiveness, lack of perceived urgency, or other factors ignored evidence of a potential threat until the looming of the stopped vehicle reached the accelerated looming stage. This stage triggered a reactive, perceptual-motor response in a last-moment attempt to avert a collision. The late stage of looming and its proximity to collision elicited a visceral, perception-action sequence not susceptible to a deliberate, decision-making process. This purely reactive response to accelerated looming is qualitatively different from that of the looming sensory threshold stage where there is usually time for more deliberative driving behaviors before the collision.

In the example of looming given in Figure 3.2, the period described by Markulla et al. (2016) and, what might be described as the point in the accelerative looming period that elicits a driver response, would depend on the target vehicle size and driver vehicle speed. In the case of the driver vehicle speed of 30 mph, this period begins for the small motorcycle at 2.3 sec prior to collision. For the case of the passenger car at this speed, the period would begin at 2.7 sec prior to collision and for the heavy truck at 3.1 sec prior to collision. For the three cases, the distance to collision is 66 ft (20 m) for the small motorcycle, 119 ft (61 m) for the passenger car, and 136 ft (41 m) for the heavy truck. The distance to stop at 30 mph with a PRT of 1.5 sec is 109 ft (33 m). The response threshold under optimal braking conditions can avoid a collision at the early stages of accelerative looming if it begins at a retinal size of 0.02 rad/sec (1.15 deg) for vehicles with the size of passenger cars and trucks. However, there is inadequate stopping time available to avoid a collision even with heavy braking, in the case of the small motorcycle using the 0.02 rad/sec retinal size response threshold. Even with a PRT of 1 sec and heavy braking, a collision would occur with the small motorcycle.

Evidence from other accident data sources involving rear-end collisions with stopped vehicles also points to this reaction to an imminent collision. Examinations of vehicles equipped with an event data recorder (EDR) involved in rear-end collisions also suggest that some drivers are initiating braking very late in the looming process (Kusano and Gabler, 2011). This study consisted of 47 cases each initiating heavy braking (19 ft/sec²) at TTCs ranging from 1.1 to 1.4 sec from collision. Adding a PRT of 1.5 sec to these brake initiation times would mean that the accelerative looming had begun at 2.6 to 2.9 sec before collision using the 0.02 rad/sec retinal image size. While the specifics in vehicle size were not provided, it is not unreasonable to assume that most of the target vehicles were passenger cars. In our example given above from Figure 3.2, the passenger car size used results in a retinal image subtending 0.02 rad/sec at 2.7 sec from collision.

The use of the sensory looming threshold as a visual cue to imminent collision provides the driver adequate time to respond provided the driver uses the information given. For a variety of reasons, some drivers do not reduce speed at all when the looming sensory threshold is reached. Only until the more compelling accelerative looming stage has occurred with the retinal image nearly an order of magnitude larger does the driver finally react. At this point, the driver is compelled to react, usually by emergency braking, to avoid a collision. This behavior is particularly problematic in the case of collision avoidance with small vehicles such as motorcycles and bicycles. Even at 30 mph, the driver who fails to brake in response to the initial sensory looming threshold of a small vehicle and is compelled to do so when the accelerative looming stage is reached will not be able to avoid a collision. At 60 mph, this type of driver behavior is virtually certain to result in a collision regardless of the target vehicle size.

OTHER VISUAL CUES AFFECTING TTC

Yet it is still possible that other visual cues may be involved in the process of imminent collision avoidance. As part of the task of TTC estimation, the accuracy of the driver's judgment of vehicle velocity could impact an estimate of the TTC to a stationary object. The strength of the rate of expansion cue in estimating TTC may be due to limits in other visual information available for judging *motion in depth*. Unlike the pictorial distance cues described at the beginning of this chapter, there are few cues to determine the distances of objects relative to the observer when those objects are moving or when both the observer and the object are moving together as is often the case in driving. As noted in the previous chapter (Chapter 2), the driver needs to extract this information from the optic flow of stimuli crossing the retina. Tau and the rate of expansion are two means by which a driver might estimate the movement of objects in depth and the TTC with those objects.

Other motion-in-depth cues include stereo motion depth or distance cues that result from the fact that the driver's visual system is being presented with two slightly different images as a result of the projection of these images on two eyes separated in space. Stereopsis is the cue to depth or distance which results from this binocular disparity in the projection of the two images to the brain. However, stereopsis normally occurs at much shorter distances than those that typically occur in driving tasks described thus far. Distances involved in the avoidance of collisions are on the order of tens or hundreds of feet depending upon vehicle speed.

Tau and other variations based on the rate of expansion of the driver's retinal image appear to be strong cues to TTC. A review and some experiments of tau and its derivatives in the presence or absence of other potential visual cues for TTC estimates suggests that these cues play only a minor role in TTC estimates (Yan et al., 2011). In this study, different aspects of the target image were manipulated such as shadowing and variations in target size, distance to collision, and speed. Once again TTC based on object retinal size expansion proved the predominant cue.

BRAKING CUES

In the situation where the driver is following a vehicle and the lead vehicle applies braking, the driver has been provided with reliable visual cues that the vehicle ahead is slowing. However, a brake light does not inform the following driver as to the *amount* of braking only that brake has been applied. Indeed it is likely that drivers assume that routine light braking is happening since that is the most common occurrence on the roadway. Heavy or emergency braking which rapidly decelerates the vehicle must rely on other cues. Visual cues for the estimate of TTC, such as tau, may influence not just when braking occurs but how braking is being controlled. Braking that is initiated prior to, or as a result of, the looming sensory threshold can be controlled by the use of visual cues other than retinal size changes. These include those involved in the control of self-motion that is part of optic flow. Evidence for the use of optical expansion of the lead vehicle and other optic flow information to the following drive is provided by a study by Liebermann et al. (1995) and by DeLucia and Tharanathan (2009). The optic expansion was found to be a visual cue to modulate braking control by the following driver. The onset of brake lights, however, triggered a strong, reactive braking response, particularly at shorter following distances.

NIGHT AND FOG DRIVING

Thus far the evidence for the use of visual cues in TTC estimation and braking control has been limited to photopic conditions. The reduction of luminance to mesopic levels has significant effects on driver estimates of TTC. When within headlight range, TTC estimates will likely be comparable to those under photopic driving conditions as the visual angle subtended by the target vehicle will be the same. However, if the target vehicle is beyond the range of the driver's vehicle headlights, the visual angle will be calculated based on the lead vehicle's observable features. In the case of most vehicles, this will either be the distance between headlights, in the case of approaching vehicles, or the distance between tail lights in the case of a lead vehicle. Evidence for this has been found in a study which manipulated the distance between the headlights of an oncoming vehicle to measure the impact on TTC estimates (Castro et al., 2005). Approach vehicle distances varied between 197 ft (60 m) and 2854 (870 m). Participants were required to estimate the distance of the approaching vehicle traveling at 36 mph (60 kph). With a width of 1 m between headlights, the drivers tended to *overestimate* the distance of the approaching vehicle. That is, they perceived the vehicle as being further away than it actually was. When the distance between headlights was increased from 3.3 ft (1 m) to 6.6 ft (2 m), the approaching vehicle distance was underestimated, appearing to be closer than it actually was.

A similar phenomenon is found when drivers follow lead vehicles in dense fog. In this case, only taillights can be seen and the distance between taillights becomes the primary cue to distance (Cavallo et al., 2001). Once again when only one taillight was illuminated or when the lights were close together, drivers overestimated the following distance. In this study, estimates of distance from a lead vehicle ranging from 26 to 92 ft (8 to 28 m) were much closer than that distance. These findings are similar to those found for motorcycles under mesopic conditions (Gould et al., 2012). Estimates of the approach speed of motorcycles equipped with a single headlight were less accurate than for cars. The addition of a tri-headlight improved the accuracy of motorcycle approach speed estimates by drivers. This is likely due to the added distance between headlights that such a design allows.

FAMILIAR SIZE AND PERCEPTUAL SET

Through experience, the human observer learns to associate familiar objects with specific retinal image sizes at a given distance. Thus, the vehicle's perceived size correlates with its estimated distance from the driver. Estimate of the distance of a specific vehicle from the driver depends in part upon how it is classified as well as its perceived retinal size. *Perceptual set* is a tendency to classify perceived objects as belonging to the most common set of objects occurring in that environment. In the case of the unusually small motorcycle in the above example, misclassification of it as belonging to the much more common passenger car class may influence how the driver estimates its time-to-arrival as seen in the SAE phenomenon. This is a particular problem when classifying pedestrians at a distance. A child pedestrian may be perceived as an adult due to the perceptual set and the smaller perceived size interpreted as an adult pedestrian that is at a greater, overestimated distance. Overestimation of the distance of child pedestrians has been confirmed through analyses of national pedestrian accident data (Stewart et al., 1993).

DRIVER EXPERIENCE

In a study of actual driver TTC estimation, both the size of the visual field and driver experience affected the TTC estimate (Cavallo and Laurent, 1988). As for experience level, the use of available visual information in making TTC estimates may be, for a variety of reasons, more accurate for the experienced driver. This includes a much better and broader visual scan of the road environment than is typically the case with novice drivers. The use of optic flow information is likely to be more efficient and effective as a result. Experienced drivers may also develop strategies that use available visual cues more effectively than less experienced drivers. Experienced drivers, having available a larger effective field of view, improved the accuracy of that estimate. The size of the visual field affects the flow field available to the driver. That, in turn, affects the driver's ability to judge self-motion.

COLLISION AVOIDANCE RESPONSE

Thus far, the discussion of collision avoidance has centered on the collision hazard and the driver's ability to estimate TTC. As the concern here focuses on imminent collision hazard, the response must be immediate. In general, the research literature has concentrated on the braking response rather than maneuvering in avoidance. For that reason, braking to avoid collision will be addressed first.

The braking response can be divided into two categories: those that occur when the driver is fully attentive to the collision hazard and those responses that result when the driver is distracted by some task or event or is incapacitated in some way. Reviews of braking have been conducted by Green (2000) and Summala (2000) and have used a variety of categories to describe the braking response when the driver is attentive to the avoidance task. The term *expected* is descriptive of the case when the driver is fully attentive to the task and anticipates the likelihood of a braking event in the near future. Green (2000) has calculated the median brake reaction time to be between 0.7 and 0.75 sec which includes 0.2 sec for foot movement for these expected events.

The heavy braking associated with a response to an imminent collision appears to be a reaction to the summation of two critical visual cues which elicit a high level of driver urgency in response (Regan and Hamstra, 1992). The first cue is the TTC with increasingly short TTC periods associated with the accelerated looming stage of signally imminent collision. The second is the rate of expansion of the perceived vehicle, pedestrian, or other road object. The increasingly rapid rate of expansion associated with the accelerated looming stage adds to the need for urgency in response.

Using the American Association of State Highway and Transportation Officials (AASHTO) formula for stopping distance and 0.7 sec for reaction time means that the driver needs about 74 ft (22.6 m) to stop at 30 mph. In all three vehicle sizes depicted in Figure 3.2, there is adequate time provided the emergency braking response occurs at or before the accelerated looming threshold. This occurs at 100 ft (30.5 m) for the small motorcycle, 324 ft (98.8 m) for the passenger car, and 424 ft (129 m) for the heavy truck.

At the higher highway speed of 60 mph (96.6 kph), there is not enough time available to stop if the driver uses the accelerated threshold stage for the small motorcycle. About 234 ft (71.3 m) is needed for stopping at this speed while only 100 ft remains before the collision. The larger-sized vehicles allow for an earlier accelerated threshold to be used. This allows for a greater time available for stopping. Provided the 0.7 sec reaction time is used, stopping in time to avoid collision for the larger vehicles is possible. However, if the driver uses the earlier looming sensory threshold as the visual cue, which occurs at 716 ft (218 m), the driver will have adequate time to stop even for the small vehicle. This would be true for the larger vehicles as well.

Only a few studies have examined the steering response to collision avoidance. The study by Markkula et al. (2014) of rear-end collisions and near-collisions found that steering alone occurred in only 1% of the cases while steering and braking in 96%. The remaining cases had no driver response at all before collision. Driving simulator studies of rear-end collision avoidance behavior show a different pattern of response. In a study by Yang et al. (2021), a sudden deceleration of a lead vehicle resulted in a variety of brake and steering responses. Some 19% of experienced drivers used braking only, while 26% used a combination of braking and steering. When compared to experienced drivers, novice drivers used braking much more often than experienced drivers (33% vs. 19%). In another simulator study, drivers used braking combined with steering as an avoidance response on only about 12% of collision avoidance events when the lead vehicle came to a complete stop (Wang et al., 2016). The number rose to 18%, with higher deceleration rates of the lead vehicle. Finally, analyses of the decision-making of which response is likely to be made revealed that heavy braking rather than steering was more likely when the following vehicle is closer to collision (Venkatraman et al., 2016). The best perceptual predictor of the magnitude of the response (e.g., braking) in this study was a combination of optical angle subtense and tau.

As with earlier studies of collision avoidance with a lead vehicle (Adams, 1994), braking to avoid collisions with lead vehicles is the most common choice, with braking and steering combinations being the next most likely response. The use of steering as an avoidance response to collision is likely constrained by the traffic situation, particularly the possibility of conflicts with traffic other than the lead vehicle that may result from changes in the direction of travel. In any case, steering alone appears to be an unlikely choice for the avoidance of imminent collisions.

SUMMARY

Driving hazards can be segregated into two categories. Imminent hazards, specifically those involving collisions, are hazards which require immediate action from the driver. Typically, but not always, this involves braking rather than maneuvering. The second category of hazards is not imminent but may develop into imminent hazards in the absence of a driver response. These *potential hazards* will be addressed in the next chapter.

Imminent collision hazards arise when, either due to chance or the driver's failure to respond to potential hazards, the driver's vehicle speed and trajectory places the vehicle on collision course with a road object. The time-to-collision or TTC is the time that a collision will occur given that no change in velocity or trajectory of the vehicle occurs. TTC is calculated objectively as the distance from the vehicle to the object divided by time. The driver's ability to estimate TTC is a function of a variety of perceptual factors involved in distance perception including static visual cues and motion-in-depth cues that arise from the driver's optic or retinal flow field. Of primary importance in the TTC estimate is the retinal image size of the collision hazard as perceived by the driver and the rate that this perceived image size changes or expands as the hazard is approached. Also important is the role of the familiar size of objects which biases the perceptual response to some objects in favor of the more common. Estimates of TTC are, however, routinely underestimated by drivers and this underestimate increases with increases in TTC. However, TTC is sometimes overestimated resulting in the perception that there is more time available for collision avoidance response than there actually is. This appears to be a product of misperception or misclassification of objects that are of unusually small size such as child pedestrians. This misperception appears to be due to the perceptual set phenomenon where objects are misclassified as members of a more common set of objects.

The collision avoidance response occurs as a response to the looming or increasingly rapid expansion of the object's retinal size as the distance between the driver and the object is reduced. The looming of an object begins at the looming sensory threshold which occurs when the object's retinal size change reaches 0.003 rad/sec. This threshold varies with the size of the object. This is the first visual cue to the driver that the distance to an object is beginning to close. The driver's response or lack of it at this stage of looming determines whether or not the more urgent second stage of looming occurs. The second looming stage or accelerated looming threshold begins at around 0.02 rad/sec rate of looming depending on the vehicle size and speed. At this stage, the vehicle and the hazard are closing the distance at a very high rate. Emergency braking may or may not avoid a collision depending on the factors such as driver vehicle speed and the size of the collision hazard.

NOTE

1. At the time of this writing, the State of California Driving Manual requires a 3 sec following distance.

4 Hazard and Risk Perception

The separation of driving hazards into those that require an immediate response from the driver and those that may require a response in the future has been done to emphasize the different perceptual and cognitive functions that are at work in the two hazard categories. The previous chapter dealt with the first category or *imminent* hazards. Typically, these hazards have a time-to-collision (TTC) of about 4 sec or less. The TTC of imminent hazards limits the driver's time to recognize and respond to the hazard to a level that may be described as reactive in nature. While the driver might respond with braking or steering to avoid the hazard, the predominant response is braking or braking and steering combined even if steering alone might be more effective. The limited time involved in avoiding collision or loss of control precludes the more contemplative, decision-making process that a driver might prefer and is more appropriate to the second hazard category.

The second category of hazard is the one which will be discussed in this chapter. *Potential hazards* are those situations, objects, or events that may develop into hazards in the future but as yet are not imminent. Such hazards may or may not develop into imminent hazards so the driver needs to decide whether a response is needed or continued monitoring of the potential hazard will suffice. A response to the potential hazard such as a reduction of speed or change of direction to mitigate the hazard potential follows from the driver's decision that active monitoring is no longer sufficient. Unlike imminent hazards, potential hazards engage and demand higher levels of driver resources, perceptual and cognitive, over longer periods of time.

Before discussing the hazard perception process in more detail, it is important to identify a component of human behavior that is a part of any discussion of driving hazards. Implicit in the concept of hazard is that all hazards contain the risk of harm to all elements, human and non-human, involved. The perception of risk that a particular hazard may have for a driver, whether it is large or small, plays a role in how that driver will respond to the hazard. For this reason, a separate discussion of risk perception is provided at the end of this chapter.

POTENTIAL HAZARDS DEFINED

In order to design road systems, signage, traffic signaling, driver training programs, and other factors affecting the use of the road environment,

DOI: 10.1201/9781003454373-4

driving hazards need to be objectively defined. Various means have evolved in an attempt to define hazards that a driver might encounter, but the most common components of these hazards involve traffic conflicts which can lead to collisions between vehicles and vehicles and pedestrians, road surface irregularities, and contaminants which damage the vehicle or result in loss of control, road construction hazards including vehicles and personnel both a collision hazard, and the problems of low light levels which exacerbate the problem of hazard detection.

Not all potential hazards are visible to the driver at any given time. Some potential hazards are hidden from view due to being obscured by other vehicles, structures, or foliage, particularly near intersections. Under these conditions, heightened driver hazard perception is elicited by the awareness that this particular environment is inherently hazardous due to the reduced ability to detect a hazard should it exist. Vehicles or pedestrians which emerge from these environments often do so with little warning.

Some hazards may be partially visible but exhibit certain behaviors which signal a potential hazard. Pedestrians between parked vehicles but facing the roadway suggest the possibility of entering the roadway. The driver viewing this potential hazard will only detect it if the driver has sufficient knowledge of how pedestrians behave in order to make a *behavioral inference* as to what the pedestrian might do in the near future. Similar cases involve pedestrians at the entrance of crosswalks or who have entered a crosswalk against the traffic signal. Other behavioral inferences may apply to drivers who exhibit aggressive or even reckless driving behavior and represent a potential hazard to others simply due to their style of driving. This includes following too close, failure to signal lane changes, driving at excessive speed, weaving in and out of traffic, and similar behaviors. Drivers observing this behavior should be capable of inferring that this particular vehicle represents a potential safety hazard.

Some driving tasks expose the driver to potential hazards more than others. Collision hazard increases in proximity to other vehicles in tasks such as parking, following, or overtaking maneuvers. Misjudgments of distance from other vehicles in these maneuvers are the common cause of collisions, but the failure to appreciate the potential for collision is evident by drivers' use of unsafe distances in following, overtaking with inadequate time to avoid approaching traffic, and poor proximity judgments in parking.

The broad spectrum and complexity of potential hazards means that multiple skill components are necessarily involved. Each of these components makes demands on the driver's perceptual, perceptual-motor, and cognitive resources in order to detect, classify, and respond appropriately. The demand for these resources can be mitigated by improvements in roadway design, road signage, and other factors. They can also be reduced by a variety of vehicle aids already available or under development. However, the most important factors affecting the driver's skill in the near term will be the training of driver hazard perception skills and the testing of the driver's ability to deal with potential hazards. The effectiveness of these design and training improvements and the role of driver experience on hazard perception will be addressed in the remainder of this chapter.

THE HAZARD PERCEPTION PROCESS

The process of hazard perception begins with the driver's visual scanning, search, and identification of the potential hazard. Associated with this process is the driver's ability to determine which of the elements within the driver's visual scan and search capability actually represents a hazard and which of these do not. Once identified, the driver needs to maintain an awareness of the presence of the hazard until such time that it is no longer a threat. The latter part of the hazard perception task requires continuous monitoring or awareness of the situation that presents itself. Each of the process components involved in hazard perception necessarily consumes the limited perceptual and cognitive resources of a driver. How the driver manages these resources will determine how successful the driver will be in addressing road hazards.

VISUAL SCANNING AND SEARCH

The key element of the hazard perception task is to identify the hazard in enough time to allow for driver response. This is not always a vehicle control response, such as deceleration or maneuvering, but also the monitoring response which allows the driver to withhold a vehicular response pending changes in the urgency of the potential hazard.

In earlier chapters, driver visual scanning of the roadway in both straight and curved sections was discussed in the context of vehicle lane position and speed control. The role of visual scanning was also discussed in Chapter 3 with respect to the onset of an imminent hazard where the lead vehicle slows or stops suddenly.

Visual detection of potential hazards depends on the ability of drivers to scan and search the road ahead at distances that allow the driver to detect and locate the hazard potential. Unlike the response to imminent hazards discussed in the previous chapter, the responses of experienced drivers to potential hazards are not reactive but anticipatory. An *anticipatory response* is a response that readies the driver for a future vehicular response such as braking or maneuvering to avoid the potential hazard devolving into an imminent one. The response may involve, for example, physical movements such as moving the foot off the accelerator pedal and readying it for brake use, repositioning hands on the steering wheel, and a general heightening of visual attention. The latter is perhaps the most important aspect of the anticipatory response and will be an enhanced attentiveness to the roadway and away from non-driving tasks.

The routine driver visual scanning and search activity prior to the anticipatory response stage generally has limited lateral movement of the eyes. Although lateral eye movements are limited to about ± 45 deg off the foveal axis, most eye movements occur within ± 30 deg off the foveal axis of each eye. The instantaneous coverage area is about 180 deg horizontal for both eyes combined in the absence of head movement. For this type of straight roadway, the concentration of scanning tends to be in the center of the roadway with occasional scans to either side of the road. Scans that sweep to either side of the roadway are typically indicative of areas or objects of interest or of potential hazards such as a vehicle about to enter the road. Visual fixations typically from 200 to 300 msec or more in duration on these potential hazards reflect the increased processing time needed to identify and classify the object (Kasneci et al., 2015). These fixations contrast with the saccadic movements of eyes during routine scanning with fixation points well under 1 sec in duration.

The coverage area of the macula or central visual field is an area totaling about 5 deg for each eye. (Note that only the very central 1 to 2 deg of the eye are evaluated in most driver visual testing.) This is the visual field area where fixation and identification of a potential hazard will occur as it has the highest concentration of visual receptors. Also shown is the area just outside of the foveal field extending from 5 to 18 deg off the foveal axis, parafovea, and perifovea of the eye. The area covered is larger than the macula but is still able to resolve objects of a foot or more in diameter at highway speeds without difficulty despite significant loss of acuity when compared to the macula. This area when combined with the macula covers a total of 36 deg for both eyes combined. The limited extent of the high-resolution areas of the eye reinforces the importance of driver visual search of the roadway in hazard perception. Beyond 18 deg off the fovea axis is the visual periphery. Evidence for object motion detection as far as 90 deg off the foveal axis has been found (Monaco et al., 2007).

THE SEARCH FOR HAZARDS

Drivers can begin the search for potentially hazardous road conditions even before beginning a trip. This strategic level of search for potential hazard conditions identifies the areas of the road system that are known to be accident-prone or under construction or are susceptible to loss of control under adverse conditions, etc. The strategic level of hazard avoidance can be complemented by a tactical level of avoidance strategies conducted during the trip itself. These are particularly useful for avoiding traffic congestion areas or areas where traffic accidents have occurred or emergency vehicles are in use as these are often broadcasted during the trip. However, our main concern here is with the operational-level avoidance of potential hazards. That is those road hazards, visible or hidden, which come to the driver's attention and require monitoring or an avoidance response during actual vehicle operations. The objective of the driver is to reduce the likelihood that the potential hazard becomes imminent.

Potential hazards at the operational level require a visual scan and search of sufficient distance from the vehicle so as to allow the driver time to visually attend to the potential hazards and to respond, if needed, to mitigate the hazard. In the previous chapter (Chapter 2), it was noted that the sight distances used for design purposes typically use a perception reaction time (PRT) of 2.5 sec with the assumption that this response time in conjunction with the actual stopping time of the vehicle forward progress would be sufficient for the driver to effectively avoid *imminent* hazards such as a stopped vehicle in the roadway. This necessarily involves a level of urgency on the part of the driver in the braking process. As was noted in Chapter 3, this level of urgency appears to occur when the rate of the vehicle image expansion begins to accelerate. In the case of potential hazards, however, this level of urgency is normally not present as such hazards are not, by definition, close enough to elicit an urgent driver response. Instead, the driver's response must be more measured, allowing for the fact that an overly aggressive response to a potential future hazard may itself create an imminent hazard as a result.

Eye movement analyses of drivers exposed to potential hazards have been used as one means of determining how drivers, novice and experienced, search the roadway for hazards. It was noted earlier that drivers will focus their fixations almost exclusively on lead vehicles if those vehicles are within 2 sec or less of the following vehicle. This pattern of eye fixations suggests that the driver considers that the lead vehicle is a potential hazard. This prioritization substantially reduces the driver's visual attention to the other areas of the road. However, when there is no lead vehicle or the lead vehicle is far beyond the range of a potential hazard, the driver is free to expend more search effort on other areas of the roadway. The precise distance of this search from the driver is likely to be determined not only by the driver's current speed but also by the driver's knowledge of how the length of time needed to detect and classify the threat level of the hazard impacts their ability to respond.

Studies of driver eye movements allow the process of hazard search to be examined in more detail since the recordings reflect both saccadic movements and fixations. A driving simulator study of eye movement patterns in straight road segments with no other vehicles or pedestrians present allowed the driver to choose fixations when no potential hazards are present (Rogers et al., 2005). The study revealed that drivers concentrated visual attention primarily at the point where the roadway met the horizon. This point is coincident with the focus on the expansion of the optic flow field discussed in Chapter 2. The concentration of visual fixations in this area increased as vehicle speed increased from an average of 18 mph (30 kph) to 114 mph (190 kph). As a result, far fewer fixations were found for areas along either side of the road. The drivers in this study focused their visual attention solely on a distant focal point, which provided steering guidance. As speed increased, the demand for that focal attention increased as well, resulting in further restrictions of visual fixations only to this area.

When more road complexity is introduced into the driving environment, experienced drivers shift visual focal attention away from the support of steering guidance and increasingly to the surrounding road environment. In a driving simulator study by Mackenzie and Harris (2015), the eye movements of experienced drivers were recorded for suburban and urban environments. An image of a fully developed collision on the road ahead was presented at random intervals to the driver and saccade, and fixation times were analyzed. The onset time of the hazard to the time the driver fixated on the hazard averaged about 2.4 sec. The elapsed time from fixation to a driver response (in this study a button press) averaged about 3.2 sec. The average time for a driver to respond to this particular hazard was thus about 5.6 sec. This would translate to an elapsed distance of 246 ft at 30 mph from the time of hazard onset to the driver's confirmation that the image was a potential hazard. This does not include the time required for a vehicular response. (Note that the PRT of 2.5 sec is the 85th percentile of the distribution of braking response times as opposed to the arithmetic mean used in this study.) This study is suggestive of much longer reaction times to hazards in complex road environments due to increased visual demands.

Previous studies using video clips of roadways indicate that drivers scan ahead of the vehicle more than 2 sec when there is no vehicle ahead while alternating between near (about 1 sec ahead) and far ahead (Underwood et al., 2003). This was also found in an earlier study where the drivers' visual scan patterns were revealed to move up and down the roadway ahead of the vehicle in the absence of a lead vehicle (Liu, 1998). The proportion of time spent looking ahead when there is no lead vehicle is also reduced by about half and the horizontal spread of the search is reduced by about 10% (Crundall et al., 2004). In the following task, typical of heavy urban commuter traffic, the driver's visual and attentional resources are dominated by the vehicle following task. In the case where there is no following task, the visual and attentional resources are freed to deal with other potential road hazards such as vehicle and pedestrian road incursions or potential hazards further down the roadway.

A question remains as to how far ahead of the vehicle drivers scan when not following other vehicles.[1] The answer to this question depends upon the driving environment, but normally a scan is more than 2 sec ahead of the vehicle and as far away as the horizon. Driver speed on straight, level roads absent traffic likely determines the concentration of focus on areas out to the horizon with broader distribution of visual fixations at slow speeds and a narrowing of the distribution at the focus of expansion as speed increases.

While the driver needs to scan ahead of the vehicle in relation to the vehicle speed, scanning laterally is necessary to identify hazards entering the roadway. Since eye movements laterally are typically only about ±30 deg off the foveal axis, the immediate horizontal sweep is only able to concentrate the high-resolution 5 deg per eye central field, which contains the highest density of visual receptors. As nearly 60% or more of the resolving power of the eye is lost beyond 5 deg of the central visual field with 80% loss at 10 deg. This loss drops further to 94% loss of acuity at 30 deg off-axis (Strasburger et al., 2011). While the visual acuity of the driver is significantly reduced when the eye is not fixated on the target, motion detection of objects is possible well beyond the central visual field. However, the threshold for motion detection will increase with increases in eccentricity.

ROAD ENVIRONMENT AND HAZARD DETECTION

Road environments vary widely in complexity depending on where they are located. Rural, suburban, and urban road environments present a variety of potential hazards to the driver. Rural hazards include many hazards rarely found elsewhere including large animals on or entering the roadway, slow-moving farm vehicles, and hilly and severe road curvature in mountainous areas, to name just a few. The hazard risk is amplified by the higher speeds typical in rural environments where sight distances may be severely compromised by terrain. In suburban and urban environments, vehicle speeds are much lower but the frequency of potential conflicts with other vehicles and pedestrians is much greater. The hazards in these environments are largely, but not exclusively, traffic and pedestrian conflicts which occur under varying speed conditions.

A study of the impact of five different road environments on hazard detection has been conducted by Beanland and Wynne (2019). The study compared the identification of critical hazards in the city, suburban, urban, rural mountain, and high-speed motorways. Critical hazards were defined as visible potential hazards such as pedestrians entering the roadway. The performance of drivers in identifying these hazards was best for city environments despite the high density of traffic pedestrians and roadside clutter. The next best environments for hazard identification were suburbs and urban environments, with mountain-rural and motorways being the worst for driver performance. A common predictor of performance across these widely disparate environments was vehicle speed. The road speed limits were lowest in the city, increased with suburban and urban, and were highest in the mountain-rural and motorway environments.

The effect of vehicle speed on hazard identification suggests that the time available for visual scans as well as the processing time required to identify the target as a hazard plays an essential role in hazard perception. It is not surprising that the lower speeds of a road environment allow more time available to perform the hazard perception task. Nonetheless, even experienced drivers may miss hazards. For the Beanland and Wynne (2019) study, experienced drivers averaging more than 17 years of experience identified only 44% of all visible hazards in the study.

The more complex the road environment is with regard to potential hazards, the greater will be the visual and mental workload on the driver (Edquist et al., 2012). The driver is forced to decrease fixation time for individual potential hazards in order to devote more time to searching (Chapman and Underwood, 1998). The performance of drivers in a hazard perception task will be affected by the number of hazards that occur within a given traffic scene.

Eye movement studies of hazard perception in complex traffic and pedestrian environments reveal that each potential hazard fixation reduces the likelihood of another nearby hazard being dictated. This phenomenon has been termed *subsequent search misses* or SSMs (Cain, 2013). SSMs have been found in the driving studies of hazard perception where multiple hazards are presented together. In a study by Sall and Feng (2016), drivers who detected highly salient hazards subsequently missed a less salient hazard. Similar findings were found in another later study by the same authors (Sall and Feng, 2019). Multiple hazards in the same traffic scene tax the visual search and processing capabilities of drivers to the extent that multiple hazard detection is compromised. It is expected that these driver limitations in perceptual and cognitive processing would then be compromised further by increased vehicle speed.

ROAD GEOMETRY AND STRUCTURES

In normal traffic patterns, even in the absence of a lead vehicle, visual scan patterns are affected by the demands of road geometry as well as by other complexities of the road environment. In straight road segments, the driver visual scan moves forward and back along the road. The central visual field fixations remain largely in the center of the road while the visual periphery is used to maintain lane position. Only when the geometry of the road changes does the demand on the driver's visual scan change significantly with regard to the positioning of the central visual field.

As road curvature becomes more demanding, the driver will increasingly focus fixations on the area of the road that provides the clearest indication of road curvature. As described in Chapter 2, this is most often the tangent point of the curve typically provided by road lane or edge markings.

The more demanding road curvatures are those with a radius under 328 ft (100 m). These curves often represent a hazard to drivers due to their high demand on driver steering performance but also due to driver failures to reduce speed to accommodate the curvature. This may result in loss of control and road departure. However, the more demanding curvatures necessarily demand greater visual attention to the road curvature and thus less visual attention to the rest of the road environment. As the focus of the driver is on the tangent point of the curve for steering guidance, the remaining portion of the field is restrained to that area of the road. This is confirmed by a field study of driver fixations in curves (Peng et al., 2018). In the study, the distribution of visual fixations ranged from 0 to 30 deg horizontally in the direction of the turn. The preponderance of visual attention was closer to the vehicle and to the lane and road edge boundaries, leaving only peripheral vision for hazard detection. The substantially lower resolution in the visual periphery impairs hazard detection, particularly for those objects that have minimal or no movement.

As the demands of road curvature increase, the ability to search and detect other road hazards decreases proportionally. Potential road hazards, particularly objects of low saliency or size such as bicycles and pedestrians, are less likely to be detected. The risk is greater for those objects that are furthest from the driver's focal point of visual attention.

Studies of responses of drivers to road structures such as guard rails and lane barriers are revealing what drivers perceive as potential hazards. For example, drivers tend to shift their lane position away from guard rails and safety barriers and then slow down (van der Horst and de Ridder, 2007). This occurred in the study regardless of the size of the barrier of the guard rail. Trees along the side of the road did not elicit the same behavior unless within 6.5 ft (2 m) from the road edge; the effect was eliminated if the trees were placed 14.8 ft (4.5 m) from the road edge. Similar results have been found by Calvi (2015). These findings appeared at highway speeds and suggest that drivers are adjusting clearance margins to allow for larger lane position variations to avoid potential collisions with fixed road objects. This also applies when overtaking bicycles on rural roads that lack lanes for bicycle traffic. The magnitude of the driver's avoidance response to these potential hazards has been shown to be affected by vehicle speed and the presence of approaching vehicles as shown by Chapman and Noyce (2014).

VISUAL ATTENTION

Thus far, the process of hazard perception has focused on the visual scan and search of the driver within the road environment. This might be described as the sensory component of the hazard perception process. The driver cannot detect a hazardous object or situation without first finding and fixating it. The next step requires the driver to *attend* to that visual fixation. That is, the available physical and mental processes need to be allocated to the task of identifying and classifying the target as a potential hazard or rejecting it as such. This visual attention sets in process higher-level processing, particularly the processing of image attributes and characteristics of the target. These are matched to internally stored images of similar target characteristics. Fixation time on a target is an indicator that visual attention is being directed to that target. If sufficient evidence is found that the externally perceived target image is a hazard, then an appropriate monitoring or avoidance response can be made. (This response is also dependent on risk criteria as discussed below.)

The mental effort during the fixation period impacts the size of the useful field of view (UFOV). The UFOV under low foveal task loads allows more attention to be allocated to the visual periphery. Experimental studies have demonstrated that the UFOV will contract in angular size as the degree of visual attention in the fovea (focal attention) increases (Ball et al., 1993). This results in slower response time or more misses to targets outside of the foveal field than might otherwise occur under a lower foveal workload. This contraction of the UFOV varies not only as a function of visual attention in the foveal field but also due to factors such as training and experience discussed below and aging (see Chapter 6).

Attending to potential hazards or hazardous situations is a normal function of the experienced driver. This is a deliberate allocation of physical and mental resources to a situation that presents risks to the vehicle. However, not all hazard detection processes are initiated by the driver. Studies of driver eye movements show an *attentional capture* process for vehicles entering roadways for both novice and experienced drivers (Underwood et al., 2005). While object movement attracts attention, other object characteristics are also effective. These include highly contrasting colors, which make the object stand out from its background and lighting, including running lights on vehicles or other road objects whether day or night.

Attention to the target, whether volitional or not, is only the first stage in the recognition of a potential hazard, however. A shift or saccade of eye movement follows the capture. This saccadic movement can be as fast as 500 deg/sec and results in the placement of the high-resolution, central visual field on the object of interest. Once the object is visually fixated, the shift of focal attention on the object must be of sufficient duration to allow the driver to process the object and classify it as a hazard. Roadsides, particularly in urban environments, are often cluttered with signs and other objects, fixed or moving, often intended to attract driver attention. To the extent that these distractions are effective, they compete with legitimate potential hazards for the attentional resources of the driver. In contrast to the central visual field, the peripheral detection of potential hazards is likely to be much less effective due to the longer time requirements for saccadic movements of the eye and the need for object movement and salience that supports the peripheral detection process. A more efficient and effective strategy for the driver is to initiate a visual search process for potential hazards based on the level of threat and the likely location of these threats in the roadway.

HAZARD RECOGNITION

The last stage of the hazard perception process is the actual recognition of a potential hazard. This is a process of matching attributes of the perceived potential target with those within the driver's long-term memory storage of similar target attributes or characteristics. This retrieval likely depends on holding the potential hazard image in working memory until a response is required. This is confirmed by data showing that drivers with poor working memory capacity have a reduced capability to recognize road hazards (Wood et al., 2010). This *classification* process likely accounts for the largest portion of the target fixation period and, along with focal attention, the greatest expenditure of driver cognitive resources.

Unlike the response to an imminent collision hazard, which is a reaction to a strong looming stimulus, the identification of a potential hazard usually involves more complex, higher-level, cognitive analyses. This is especially true for hidden hazards which require the identification of precursors or clues that may portend a hazard.

For the identification of visible hazards, the driver needs to understand what attributes of objects or situations represent a potential threat of collision to the driver's vehicle or the loss of control of that vehicle. The potential hazard must fall within a specific distance of the vehicle traveling at a given speed to represent a threat. The distance of the threat determines the time available for the driver to respond. The time available then determines whether the driver will continue monitoring or will engage in some type of vehicular response.

The recognition of a potential hazard requires the knowledge of driving situations or scenarios that give rise to potential hazards as well as the hazards themselves. Thus, driver training and driving experience is required to build a knowledge base of potential hazards. Testing of hazard perception ability is the first step in determining whether a driver has the knowledge available to recognize and respond to potential hazards.

HAZARD PERCEPTION TESTING

Testing hazard perception ability in novices and experienced drivers occurs in two forms. In one form, a series of static or video clips of potential hazards are presented with appropriate non-hazardous lures. The latter are provided to determine driver false alarm rates. The response of the driver is a simple button press. Typical metrics include the speed of response of the driver and the driver's accuracy in locating potential hazards in a given scene. More recently, eye movement analyses have been added to the list of metrics of hazard perception in these studies. Additional methods of testing include the use of driving simulators and real-world driving, although the use of static images and video clips remains the most commonly employed of the methods available.

A review of 49 studies employing these various methodologies was conducted to evaluate their efficacy in assessing the hazard detection performance of groups of drivers varying in age and experience (Moran et al., 2019). The review found that all the major methodologies employed in these studies were able to differentiate hazard performance between age- and experience-level categories on at least one measure of performance—response speed. Findings were inconsistent, however, on the metric of accuracy in locating hazards. Inconsistencies were also found in how investigators categorized drivers by age and experience. For example, age and experience are correlated with the perception of risk, particularly in males. The relationship of the risk level associated with driving hazards will be discussed below in the section on *Risk Perception*.

Tests of the ability of novice drivers to detect hazards show relatively poorer performance when compared to experienced drivers. In a study of response time to detect a hazard in static image tests, novices were found to be significantly slower in response than experienced drivers (Quimby and Watts, 1981; McKenna and Crick, 1991; Borowsky et al., 2010; Wetton et al., 2010). Testing of novice drivers with as little as 1 month of driving experience compared to experienced drivers using video clips of hazards showed longer response times for novices, but the difference was not statistically reliable (Sagberg and Bjornskau, 2006).

In a static test of driver experience levels, novices responded more slowly to potential hazards and rated traffic conflicts as less hazardous than more experienced drivers (Scialfa et al., 2012). In a test using video clips, experienced drivers identified more hazards than novices, regardless of road type (Beanland and Wynne, 2019). Finally a study of drivers used probe questions to determine the level of situation awareness of the driver during hazard perception (Crundall, 2016). This testing method also reliably discriminated between novice and experienced drivers.

The second form of testing is to expose the driver to potential hazards in a driving simulator or in a real vehicle and allow the driver to respond to the potential hazards with a vehicular response such as braking or maneuvering as they would in the real environment. In one such driving simulator study, young-novice drivers' hidden hazard detection was compared to that of older-experienced drivers. Detection was assessed by recording the presence of visual fixation on the hidden hazard. The novice drivers performed more poorly than experienced drivers in the detection of hidden hazards (Pradhan et al., 2011). Another study using a driving simulator hazard test found no differences between novice and experienced drivers in response time to a potential hazard detection (Hirsch et al., 2015). However, a composite measure of TTC (C-TTC), which calculates the TTC moment by moment within a specific visual angle, did show reliable differences between novice and experienced drivers in this study. The C-TTC measure is sensitive to the anticipatory speed changes that differentiate the novice from the experienced driver. A more recent driving simulator study found that novice drivers were *faster* in detecting hazards than experienced drivers even though both groups had the same hazard recognition performance (Borhan et al., 2019). Driving frequency and experience with different road environments may account for some of the results.

This brief review of hazard perception testing was intended to reveal what aspects of driving hazard perception are influenced or affected by driver experience. The development of predictive or anticipatory skills regarding potential hazards is one of these components. A recent review of hazard perception studies generally supports the use of testing of response time in detecting potential hazards but not the use of tests based on driver accuracy in locating such hazards (Moran et al., 2019). However, the evaluation of only response time and not accuracy makes it difficult to assess false alarm rates known to be a factor in the hazard perception response. The inconsistencies in some study results suggest that a more rigorous standardization process for testing is needed for research, particularly if used as part of the licensure process. Other more recent reviews of the hazard testing literature have come to the same conclusion (Cao et al., 2022).

Standardized testing with humans, whether in driving or elsewhere, requires a well-established, systematic process of development and evaluation with large, representative population samples. The process must include content validity, test–retest reliability indices, and criterion-related validity measures. Content validity refers to the requirement that the content of the test is a valid, representative sample of the hazard perception behavior of drivers. Test–retest reliability assures that the test results are not just a function of random, temporary characteristics of the driver but a reliable estimate of driver ability. Criterion validity, a measure of the predictive power of the test, is especially important to driving safety. There should be a high correlation between driver scores on hazard tests and some criterion measures of driver safety such as accident rate. These are only a few of the measures needed to construct valid testing of driver hazard perception which are, for the most part, currently lacking in hazard perception testing.

HAZARD PERCEPTION TRAINING

The poorer performance of novice drivers on hazard perception tests suggests that current driver training programs are deficient in this area. Hazard perception as a skill is now being acquired after licensure on the road. This inevitably contributes to a higher accident rate of novice drivers until such time as the skill is developed. However, it is also evident that a normal driving experience does not necessarily result in an optimized hazard perception skill. While experienced drivers have higher hazard perception performance than novices, they still have significant hazard perception error rates.

The question then is not whether additional training in hazard perception is needed but what form it should take. A review of some of the earliest training studies on young drivers under 21 years of age has been conducted by McDonald and associates (2015). Training methods included the use of computer programs, video clips, driving simulators, and real-world vehicles. Evaluations were generally conducted either immediately or a few days after training using computer-based testing or driving simulators. All studies showed improvement in hazard perception performance for trainees.

Only one study in this review examined the effects of novice training after a period of 1 year and is thus particularly worthy of note (Taylor et al., 2011). This study used a computer-based training program using driver fixations as a performance measure. At intervals of 6 months and up to 1 year, the trainee's performance was evaluated using driver fixations in a real-world field driving test. Trainee performance in hazard perception was found to be superior to a control group even up to a year after training. While superior to the control group of untrained novice drivers, trained novice driver hazard perception was still inferior to experienced drivers.

One area of hazard perception training for novice drivers has proven difficult, however.

Hidden or latent hazards which do not appear to the driver due to obscuration by other road objects have proven to be particularly difficult for novice drivers to recognize. A review of latent hazard anticipation training studies using error-feedback training for young drivers was conducted to address this issue (Unverricht et al., 2018). The review compared the efficacy of different training modalities (computer or non-computer-based) and types of evaluation (simulator vs. on-road evaluation). The results revealed that both training modalities improved latent hazard performance to a similar degree for novice trainees as measured by driver visual fixations. The evaluation method used was effective in showing trainee improvements. The training had a greater effect on 18- to 21-year-old trainees than on younger 16- to 17-year-olds. Notably proving different perspectives of the potential hazard from both the egocentric (driver) and exocentric (outside vehicle) viewpoints was more effective than either view alone. The feature of these studies may be particularly important as a later study of latent hazard recognition using more traditional training methods did not find improvements in novice recognition of latent hazards (Kahany-Levy et al., 2019).

The effectiveness of providing different visual perspectives of latent hazards is likely the result of the construction of internal representations of these latent hazards that cannot be developed with either an egocentric or exocentric view alone. Potential hazards, unlike imminent hazards, are cognitive constructions of the road environment that develop from active processing of objects and their interrelationship. This construction process takes time and effort requiring error feedback and repetition as with any other training of a complex task. The slower response of novice drivers to potential driving hazards likely reflects the longer processing time needed to identify road hazards. However, it may also reflect the higher risk thresholds these novice drivers have for hazardous events.

RISK PERCEPTION

The discussion thus far has centered on the perception of hazards by drivers. The perception of a hazard is the detection of an object, event, or situation that might cause the driver, the vehicle, vehicle passengers, or other objects in the roadway environment some harm. The *perception of risk* is intrinsically tied to hazard perception as risk perception is about the *probability* that harm will be caused by that hazard. Thus, the driver's perception of hazard and risk are bound together.

A potential hazard can have perceived risk levels (probability of harm) varying from very low to very high depending on the particular driver's level of risk sensitivity. Many models of driving behavior have been developed with driver risk perception as a central element. Early theories of driving behavior focused almost exclusively on predicting driver behavior based on risk threshold measurements (see Summala, 1996 for a review). For the purpose of comparison, the term *risk-averse* will be used here to describe those drivers who are especially sensitive to risk while *risk-seeking* applies to those drivers who actively seek risk. Lastly, *risk-tolerant* drivers are those who recognize that driving involves some degree of risk but tolerate that risk in order to derive the benefits of driving. Naturalistic driving data have shown that low-risk (defined here as risk-tolerant and risk-averse) drivers make up the vast majority of drivers (84%) while the remainder of drivers are likely classified as high or moderate risk (defined here as risk-seeking) (Guo and Fang, 2012). Demographic, personality, and driving characteristics were used to identify these groups. The relationship between accident rates of individual driver risk profiles and risk as defined by accident data such as crashes and near-crashes, however, does not include a significant proportion of accidents that are not reported (about 30% in the U.S.). This fact needs to be kept in mind in any assessment of risk behavior and accident rates.

Individual risk sensitivity levels will impact how an individual driver will respond once the hazard has been detected. Individuals who are, for example, risk averse are likely to treat hazards as a greater threat of harm than those who are not risk averse even though the hazard is identical. Drivers who are more risk-averse are likely to be more sensitive to the road environment and traffic conditions than other drivers. In a study comparing hazard perception of urban and rural environments, those drivers with higher risk sensitivity were more likely to judge rural environments as inherently less risky than urban road environments (Cox et al., 2017). This was true even though the hazard scenarios were otherwise very similar between the two environments. Furthermore, risk-averse drivers will likely bias their response toward identifying an event as a hazard while more risk-tolerant drivers will not. Older drivers are more likely to be risk-averse drivers than any other age category (see Chapter 6). These drivers often drive at speeds below the posted limit because they associate speed with risk. They also tend to avoid high-traffic environments and freeway driving as well. These compensatory behaviors are often associated with risk-averse drivers.

Alternatively, drivers who are risk-seekers often engage in hazard seeking such as high-speed driving. The intrinsic pleasure of risky driving outweighs the possible consequences. A strong correlation has been found between young-novice drivers (less than 25 years of age) identified as risky drivers and their subsequent involvement in crashes (Deery, 1999; Ivers et al., 2009). One reason for the high accident rate of young drivers may not be just their poor hazard detection but how the hazard is processed once it is detected.

DRIVER RESPONSE CRITERION

One of the reasons that novice drivers take longer to respond to potential hazards appears to be their higher threshold for risk. Accordingly, novice drivers may perceive road hazards, but the hazards do not exceed the diver's risk threshold. As a result, a response to the hazard is either delayed or withheld entirely. Wallis and Horswill (2007) investigated this issue in a study using techniques used in signal detection theory to segregate the driver's sensory-perceptual sensitivity to a hazard from the criterion level or response bias that a driver may have to the risk associated with that hazard. The signal detection model uses hit rates, false alarm rates, miss rates, and correct rejections to derive a measure of hazard perception, d prime (d'), and a separate measure of response criterion (β). Separate scores associated with each of these elements allow statistical tests to be conducted on each.

In this study, novice drivers were trained in hazard perception using a variety of road hazards in a video-based simulation. A second novice group and an experienced driver group received no additional training. Results revealed that hazard perception sensitivity (d') showed no difference between experienced and novice drivers and no difference between trained and untrained novice drivers. Differences were found, however, in the latency of their responses and in their response criterion (β). Response bias, but not hazard detection sensitivity, was significantly correlated with response latency for all three groups in the hazard perception test. Untrained novice drivers required a much higher threshold of danger to respond to a hazard than either trained novices or experienced drivers. The higher threshold of danger resulted in longer response latencies for those untrained novices, while the trained novice response latencies were significantly faster than those of untrained novices. The authors found that novice drivers' requirement for a potential hazard to exceed a higher threshold of risk before classifying it as a hazard resulted in longer response latencies to displayed hazards.

A more recent study using a different method of hazard perception testing found somewhat different results (Ventsislavova et al., 2016). Applying the signal theory model, this study found that the d' measure of hazard perception sensitivity was greater for experienced drivers than for novice drivers. The study also found the sensitivity to be higher for those with no prior traffic offensive than those drivers who had prior offenses. A questionnaire measuring cautiousness also found that experienced drivers were more cautious when facing hazards than novices. However, this study did not find a difference in response criterion (β) between experienced and novice drivers. This may be due to a different testing method which allowed a more conventional version of signal detection measurements to be used in this study compared to a modified version used in Wallis and Horswill (2007). Whatever the reason, these inconsistencies in findings point again to the need for rigorous development standards in test development and application.

The use of a signal detection model in segregating hazard detection from response criterion shifts is an important step in isolating the ability to detect hazard from the response criterion to the hazard. However, its effectiveness may depend on the method used in testing hazard perception. There is no question, however, that differences exist between novice and experienced drivers in their risk perception of hazards. Experienced drivers, likely due to both experience and age, tend to be more risk-averse than young-novice drivers.

ROAD DESIGN AND DRIVER ESTIMATES OF RISK

Objectively, all road hazards have at least some defined risk of causing harm otherwise they would not be classified as hazards. In fact, road signage is dependent on an objectively defined risk such as excess vehicle speed in a school zone or passing other vehicles on a blind curve. However, driver perception of risk is often at variance with objective risk, resulting in the likelihood that such road signage is often ignored. In a study comparing driver estimates of risk to objective highway hazard risk, no reliable correlation could be found (Philput, 1985). Moreover, a more recent study found that when driver risk assessments are below the objective level of risk, as in the selection of vehicle speed, driver likelihood of accident involvement increases significantly (Kokubun et al., 2004).

Increasingly, those involved in road design are realizing the need for driver input regarding the degree of *perceived* risk of any road hazard by drivers. One of the factors that is most problematic when addressing road improvements is driver *behavior adaptation*. Behavior adaptation refers to the change in driver behavior resulting from a road design improvement which is other than that intended by the designer (Ward, 2000). For example, widening roads and lane width to increase safety margins in passing may result in increased driver speed due to the perception that wider roads reduce the risk associated with increased speed. Evidence for this is provided by de Waard et al. (2004).

One question is whether this adaptation is a conscious act on the part of the driver or the result of the driver perceptual processes. These perceptual processes are below consciousness and therefore are not attended to by the driver and will have no effect on perceived risk. This hypothesis was tested by Lewis-Evans and Charlton (2006) in a study on the effects of road width on driver speed, difficulty, and risk. The comparison was made on narrow-width, control, and wide-width roadways and lanes. The narrow road produced the lowest speed, the control the highest, while the widest road did not produce speeds different from the control in the driving simulator. Later risk rankings of static road images did not suggest any conscious differences in either road width or risk. These authors, along with Summala (1988), suggest that changes in road design that reduce accident risk will not be consciously detected by the driver. The perceptual processes that affect driver behavior in response to road design changes are a result of pre-conscious perceptual processes (many of these are discussed in Chapter 2). Reductions in objective road risk are below drivers' subjective threshold until circumstances make them otherwise. Other examples where road design features such as guardrails and roadside objects such as trees impact a driver's perceptual control processes were discussed earlier in this chapter. The importance of pre-testing designs for behavioral adaptations using driving simulators would help reduce or eliminate the problem of design-induced safety problems. Such driving simulators, however, need to be high in *perceptual fidelity* in order to elicit the same driver perceptual processes as would occur in the real-world implementation of the new road design (Lee, 2017).

ROAD SIGNAGE AND PERCEPTUAL CUES

The response to hazard mitigation in roadways is usually some form of signage. Road signage of various types including electronic displays has been employed in an attempt to reduce the risk associated with road hazards. Among these are road signage used to reduce the risk of negotiating curves, particularly those with more severe curvature. In a study by Charlton (2007), the use of advanced warning signs was compared to other forms of road markings designed to alert drivers to horizontal curves and to reduce speed. Road signage was not found to be effective in reducing speed when used alone when compared to when they were used together with road markings such as chevron sight boards or herringbone road markings. Most effective in reducing speed was the use of rumble strips. The author argues for the use of perceptual cues as the most effective technique in reducing speed in curves.

The effectiveness of road signage and perceptual cues in a study of driver behavior in negotiating curves was also examined by Milleville-Pennel et al. (2007). In this study, drivers viewed video clips of horizontal curves varying from 85 ft (26 m) to 1476 ft (450 m) in radius. Drivers provided a verbal estimate of both curve risk and curvature after viewing the curves. They were also asked to turn a steering wheel to the angle corresponding to that required to negotiate the curve. Drivers made estimates either or without the presence of an international curve sign. The results indicate that the curvature of very severe bends of 262 ft (80 m) or less was underestimated along with an underestimate of associated risk and poor perception of the road bend as indicated by the driver steering wheel angle (SWA). The presence of road signage for the curves improved driver estimates but did not improve SWA estimates for negotiating the curve.

The results of these studies show the importance and limitation of driver visual perception of road curvature in negotiating curves. The poorer performance in the physical estimation of curvature as evidenced by driver steering inputs is particularly concerning for severe bends. Speed control becomes increasingly important as curvature increases. In the case of very severe bends, poor speed control can result in road departure. Anticipatory steering estimates by the driver are important in the preparation for entering a curve while last-moment steering judgments are likely to be inaccurate and unrecoverable.

The role of signage in negotiating curves or elsewhere is to alert the driver to an impending hazard. However, it is not a substitute for perceptual cues, either visual or non-visual. These cues act at a pre-attentive level of processing so they are much less vulnerable to problems associated with driver attention as is the case with the use of signage. Signage serves as an alert to a potential hazard only to the extent that the driver attends to it and processes the message. To this end, improvement in the hazard alerting function of signage is needed such as the use of electronic displays. In the case of severe bends where speed control is essential, electronic displays that can feedback on driver speed and alert the driver to speed excesses eliminate the need for the driver to shift visual attention to in-vehicle speed displays at a time when that attention is needed for the road ahead.

SUMMARY

Perception of hazards, either imminent or potential, is one of the most important driver skills. However, the perception and response to these different hazards depend on the use of different perceptual and cognitive driver resources. While the response to imminent hazards relies heavily on reactive perceptual components, such as looming, the detection of potential hazards demands more visual attention and higher cognitive resources. Visual attention demands visual search and scanning activities on the part of the driver that are acquired through training and experience. Visual search processes have been shown to be affected by the complexity of both rural and urban road environments. Once the potential is visually fixated, the driver's hazard recognition process is then initiated to determine whether the target is or is not a potential hazard. A positive recognition results in a driver response in the form of either a vehicle control response or continued monitoring. The latter case requires working memory storage and continual updating of the target status.

The deficiencies of novice drivers in hazard recognition, particularly for hidden or latent hazards, are evident. This has resulted in the development of hazard perception testing for licensure in some countries. Hazard perception tests of varying types are available although there is a lack of standardization and rigorous test development which has led to inconsistencies in results. Hazard perception training has also been attempted with a wide range of modalities available. This training appears to be successful at least in the short term in improving driver hazard perception skills. More research is needed to determine the rates of skill decay and skill retraining requirements. Hazard and risk perception are bound together as any hazard so classified is, by definition, a risk. However, risk perception is rated by the probability of harm from a potential hazard and this probability affects how drivers respond. Young inexperienced drivers have higher risk thresholds, which result in a slower response to hazards than older more experienced drivers. Although driver risk perception with respect to the response to potential hazards varies with driver experience and age, it also varies with the driving environment, particularly complexity and traffic load. In the case of road hazards such as severe road curvature, risk perception may reduce the impact of signage-based alerts and warnings on those drivers who are prone to take risks. Alternatives which depend less on driver cognitive components and more on perceptual ones appear to be more effective in alerting drivers than signage.

NOTE

1. The California State Driver Manual recommends that drivers look 10 to 15 sec ahead of their vehicle.

5 Multitasking and Attention

The driving task domain is inherently a multitasking domain. The driver must maintain control of vehicle speed and direction at all times. In this case, the multitasking occurs at the level of braking and accelerator control responses and steering wheel inputs. However, other driving-related tasks such as hazard perception process, especially collision avoidance, recognizing and processing road signage and road markings, and wayfinding and navigation must be attended to at some level during the vehicle speed and directional control process as well. Added to these are the non-driving-related tasks or *distractions*, which include passenger communications, cell phone use, in-vehicle displays and controls, drinking, eating, roadside distractions such as advertising, and many others.

A variety of sensory, perceptual, perceptual-motor, and cognitive resources are engaged in the execution of these tasks whether they are driving tasks or distractions. These resources are each limited in capacity. In some tasks, a resource may be completely consumed with a particular task and the driver will have no reserve capacity available for the performance of any other task. In other driving tasks, the driver will have a reserve capacity available for distractions. It is the particular constituent elements of a driving task that make it vulnerable to interference from distractions. An analysis of driving tasks into constituent physical and mental resources reveals the potential each task has for interference either from other driving tasks or from distractions. This demand can be categorized as sensory, sensorimotor, perceptual, or cognitive.

In the case of sensorimotor, a second task is making a demand on the driver's ability to use a vehicle control such as the steering wheel because the driver's hand is in use for another task either driving-related or not. Control actions often involve a sensory system like vision or touch so there will be demand on both physical and sensory resources.

Task demand also occurs purely at the sensory level at any of the basic senses: vision, audition, tactile, somatosensory, and vestibular. Most commonly this occurs in visual focal attention where the driver's eyes are directed to one task while a second task demand cannot be fulfilled. The driver's focal vision is not sharable between tasks. A common case is the shift between looking outside for hazards while looking inside the vehicle for an indication of speed.

The third category is cognitive. Cognitive demand and interference are the most commonly investigated in driving primarily because of the role

 DOI: 10.1201/9781003454373-5

of inattention in accidents. However, cognitive demand may occur in a variety of driving and distracting tasks. As noted in earlier chapters the higher-level cognitive processes are involved in a variety of driving tasks that involve working and long-term memory such as situation awareness and the attention required for hazard perception and other tasks. Cognitive resources are limited to serial processing. This means that only one consciously controlled process can occur at any one time.

DRIVER TASK ANALYSES

An examination of the constituent components of a driver task through task analysis will reveal the vulnerability of a driver task to secondary task demands as well as the capacity of driver resources in meeting the demands of the primary driving task. A number of task characteristics have been covered in previous chapters, but the following analyses are focused on executing these tasks in a multitasking environment.

VEHICLE SPEED AND STEERING CONTROL

In Chapter 2, the role of vision and other senses was discussed in relation to the ability of a driver to control vehicle speed and lane position. The task goal for the driver in-vehicle control is to maintain lane position (lateral) control through steering wheel inputs. In straight roads absent traffic, the experienced driver will visually scan the road with the principal line of sight forward to the horizon. This *focal vision* focuses driver resources within a central visual field measuring about 10 deg in diameter. This high-resolution area also allows the driver to focus visual attention on the path ahead and to detect relatively small objects in the roadway. The peripheral area of the visual field outside of the central foveal area allows the driver to use *ambient vision*, which is specialized for spatial orientation to maintain lane position using lane or road edge visual cues. As ambient vision is functioning at the pre-attentive level, the driver's cognitive resources can be devoted to processing data captured in focal vision.

Lane position control becomes more critical with increasing road curvature such that focal vision becomes predominantly occupied with extracting curvature cues from the road surface to support steering control inputs. The relationship between the tangent point of the lane or road edge marking is used to support steering wheel angle input with an update rate of about once every 0.85 sec. With more severe curvature, the requirement to process curvature cues comes to dominant driver attention to the potential exclusion of road hazards in the visual field. As attention is directed toward matching steering angle inputs to track tangent point changes, maintaining lane position while negotiating through the curve is more dependent on ambient vision. As ambient vision is pre-attentive, its role in hazard perception is limited to moving or high-contrast objects that can capture focal attention. The fact that the driver's visual field is directed mainly, if not exclusively, to road curvature cues will reduce hazard perception processes outside the central visual field.

With regard to experience, automatization of the relationship between steering wheel inputs and road cues, such as the tangent point, substantially reduces the impact on driver attention. For novice drivers who have yet to automatize this sensorimotor relation between steering control and visual control cues from the road surface, the impact of road curvature can be substantial. Each steering control will require the driver's attention as will each steering wheel angle adjustment. This reduces even to a greater extent the ability of the novice driver to respond to road hazards.

The control of vehicle speed, while only occasionally supported by driver speedometer checks, is largely a function of ambient vision motion cues from the visual periphery supplemented by audio cues from wind, engine, and tire noise. Speed through driver ambient visual cues is generally underestimated by drivers with the underestimation greater at slower speeds than at higher ones. This dependence on visual sensation to determine visual speed is heavily influenced by the size and composition of the optic flow field. The flow field with respect to motion perception extends about 60 deg to either side of the driver's visual field and is limited in area by the vehicle frame. The perceptible elements of the road include texturing and markings on the road surface as well as the constituent elements of the roadside environment. These elements affect the optic flow and therefore the driver's perceived vehicle speed. Asymmetric optic flow characteristics generated by differences in flow patterns between road features on either side of the vehicle have been shown to affect steering behavior as well (Kountoritis et al., 2013).

HAZARD PERCEPTION AND RESPONSE

Driver collision avoidance behavior was discussed in detail in Chapters 3 and 4. The primary task in collision avoidance is to assess the time-to-collision (TTC) of objects detected on or near the roadway. This task is dependent on the driver's ability to adequately search the road for objects that may present a collision hazard. On detecting the object, the driver must estimate its TTC based on whether the object is stationary or moving toward the driver's vehicle. Estimates of TTC appear to depend heavily on the driver's perceived retinal size of the object or the change in the retinal size of the object. If the retinal size of the object is approaching the driver at a rate exceeding 0.003 rad/sec, it is exceeding the *looming threshold*. Looming is the initial cue to the driver that the object may be a collision threat. When the looming of the object exceeds about 0.02 rad/sec rate of retinal expansion, it has entered the *accelerated looming* stage. This stage provides a strong visual cue that a collision is imminent and a collision avoidance response is urgently required.

The avoidance response to an imminent collision, either braking or maneuvering, depends on the TTC and road factors such as traffic and other obstacles which may restrict maneuvering. Driver responses to imminent collisions which have reached the accelerated looming stage are reactive and more likely to be braking than maneuvering due to the short TTC. The braking reaction time under these circumstances is about 0.7 sec compared to the average of 1 sec for less urgent braking responses.

Collision avoidance also occurs in situations of less urgency, particularly when the vehicle is in close proximity to a potential hazard. This is typical of low-speed maneuvering where the driver is attempting, for example, various parking maneuvers. Typical clearances may be only 1 to 2 ft. While retinal expansion as means of determining TTC is a part of this, the driver also needs to have an internal representation of the vehicle frame with respect to other near objects. This internal representation is similar in kind to that used by a pedestrian negotiating a crowded sidewalk where an individual must avoid colliding with other pedestrians. This personal space that surrounds an individual is similar in kind to the vehicle frame representation. The representation of the vehicle frame space is developed over time by operating a specific vehicle. Interactions with the proximal driving environment establish the vehicle space requirements for the driver. How much experience is required to develop these requirements is not known. However, failure to develop this representation can create collision risk not only in parking maneuvers but also in the task of negotiating narrow roads or alleys.

The detection and response to potential or near-future hazards differs from that of imminent hazards in that more time is available to the driver to detect the presence of a hazard, to judge its potential for harm (risk), and to decide on a response. Potential hazard detection depends heavily on visually scanning and searching the road environment for potential hazards that may lead to collision or loss of control. Detection of a potential hazard requires identification based on known attributes of hazards stored in the driver's long-term memory. This identification requires visual fixations on the target of sufficient duration to allow for that identification to be completed. To maintain situation awareness of the driving environment, the driver will need to retain the hazard in working memory until the hazard passes. This allows monitoring and updating the relative position and status of the hazard. Multiple hazards may be detected, particularly in complex, urban environments, and each needs to be a part of the extant driver situation awareness until such time as the threat from each no longer exists.

Vehicle Following

Vehicle following is a common driving task that requires the driver to maintain a safe distance from the lead vehicle. The visual distance cues include the size–distance cue that provides the driver with a perceived distance of a vehicle based on its perceived retinal size. Other distance cues such as those from linear perspective and texture-density may play a role as well. While drivers can estimate distances with some accuracy at short following distances (under 65 ft or 20 m), underestimation of the following distance is common and will increase as the distance from the lead vehicle increases.

Responses to closure on the lead vehicle depend on the presence of lead vehicle brake lights and rates of change in the following vehicle driver's retinal size. However, the lead vehicle may decelerate without the use of braking. In any case, even when used, brake lights do not denote the heaviness of braking. Thus, absent any other information, the following vehicle does not know if emergency braking is being applied when brake lights illuminate. In this case, the detection of the change and the rate of change in distance will depend largely on changes in the following driver's retinal image size of the lead vehicle. Rates of change in retinal size must exceed the looming threshold in order for a change in closure to be detected. Furthermore, in order for an emergency braking response to be made by the following vehicle, it is likely that the accelerated looming stage would need to occur. Thus, the following driver needs to be attending to the lead vehicle carefully throughout the process in order to detect these changes in lead vehicle distance and closure. This emphasizes the importance of visual attention in the vehicle-following task. Currently, following the driver's distance of 1 to 2 sec from the lead vehicle is generally not sufficient to perceive changes in closure with the lead vehicle in time to avoid collision if the lead vehicle is applying full emergency braking even if an accelerated looming stage is reached. This is especially true if the roadway is contaminated such that the following vehicle stopping distance is compromised.

In the following task, a driver may use the lead vehicle as a steering guide using focal vision if the lead vehicle is at a distance that allows for the detection of changes in direction in time to input steering control. This may occur during inclement weather such as fog or other factors which may obscure steering guidance from road cues. The change direction of the lead vehicle provides an anticipatory cue to steering angle changes necessary to avoid obstacles as well as negotiate curves. However, the use of a lead vehicle for steering guidance in poor visibility conditions may put the following vehicle at an unsafe distance from the lead vehicle, thereby increasing the possibility of a collision.

OVERTAKING

As noted, drivers may be using different techniques in executing the over-taking or passing maneuvers of a lead vehicle. When overtaking another vehicle in the presence of approaching traffic, the following is now fast with multiple tasks. The first is to maintain a safe following distance from the lead vehicle. The driver then must determine whether there is adequate time to pass by determining the distance and TTC of the approaching vehicle. The distance of the approaching vehicle is determined by combinations of retinal size and other visual cues to distance available from the roadway. If the approaching vehicle is beyond a critical range of a potential hazard, the driver may execute the overtaking maneuver. If not, the driver will continue to determine the TTC of the approaching vehicle. The TTC in overtaking is the time taken for the approaching vehicle to arrive at the lead vehicle's position. If the TTC estimate is judged to be of sufficient time to execute the maneuver, the driver will move to the passing lane and accelerate. Some drivers may choose to reduce the following distance from the lead vehicle in order to reduce the time required for overtaking. This may lead to unsafe following distances in the overtaking maneuver. As drivers usually underestimate TTC, a safety margin needs to be added by the driver to the estimate. The completion of the overtaking also requires a clearance estimate from the front of the lead vehicle. Typically, the driver would apply the rule that the overtaken vehicle's headlights must appear in the vehicle's rearview mirror before completing the maneuver.

NAVIGATION AND WAYFINDING

Drivers will develop an internal representation in memory or a *cognitive map* of a route with repeated exposure. The map is constructed of waypoints consisting of physical landmarks. These waypoints are used for turns but are also used as points to determine distances from one part of the stored cognitive map to another. During the development of the map, the driver needs to pay attention to key landmarks and store them in memory. Once the map is fully developed, the driver easily identifies the landmarks and their significance in terms of where the driver is in relation to the next turn and the destination. Most driving is spent in familiar environments where these cognitive maps are used for navigation.

In unfamiliar road environments, where drivers will not have a cognitive map, drivers must rely on other information resources. These external resources for route negotiation include maps displayed visually or aurally or by in-vehicle systems or conventional paper maps. Other resources include passengers who may be more familiar with the environment or hand-held devices such as smartphones or tablets. In any case, the driver must now shift the focus of attention away from the road environment. While aural displays and passenger communication reduce the demand for visual attention, they still make demands on cognitive resources necessary to attend to and process aural communications. Aural communications have the advantage that the driver does not have to physically orient to the source as is needed for the visually displayed information.

ATTENTION

Arguably, attention is one of the most important physical and mental driver resources. It is also among the most misunderstood. Attention, both involuntary and voluntary, is the means by which the driver directs or allocates resources to a task or a task component. When describing a particular sensory channel such as focal vision, the driver is using this channel to focus not only visual receptors but also supporting visual system components necessary to process the incoming sensory data. This requires time and effort. The length of a visual fixation on a road sign or roadway marking is needed in order to provide enough time for sensory data to reach higher-level processing centers to decode the incoming data. This is particularly true for those visual tasks that require language or symbolic processing which requires higher-level cognitive processing including access to long-term and working memory.

While visual attention is a primary resource for driving, it is not the only means by which resources are directed. Aural attention, attention directed through the sense of hearing, is particularly useful for alerting drivers as there is no need for physical orientation of the sensory receptors as there is for vision. Aural attention can also be focused on a particular signal source. This is illustrated by the "cocktail party" phenomenon in which communication of primary interest allows other irrelevant communications to be filtered out or ignored. Aural attention issues will likely become more prevalent as in-vehicle advanced systems such as collision avoidance and lane position control will increasingly rely on aural alerts and communications to avoid dependence on the limitations of visual attention.

Less understood are attention mechanisms based on sensory systems other than vision or hearing. Vibrotactile alerting systems, using the driver's sense of vibration through the steering wheel or seat, is one suggested means of alerting drivers. The use and integration of these body sense or somatosensory cues as alerting systems are less understood than those based on visual or aural sensory cues.

Apart from the sense-based attention mechanism, the driver can initiate higher-level processes voluntarily and directly. Thinking about issues other than driving is not uncommon, particularly when driving task demand is low. However, whether the task is related to driving or not, the impact of using these limited cognitive resources will be the same. Those tasks which are dependent on high-level processes, including hazard perception processes and other tasks requiring decision-making, will be denied access to these cognitive resources. Thinking about issues not related to the driving task at hand is distracted driving even if it is related to a driving issue in the future (such as in tactical navigation route decisions).

USEFUL FIELD OF VIEW

The functional field of view or what has more recently been termed the *useful field of view* (UFOV) refers to the functional size of the field of view when the driver's focal attention is fixated on an object of interest. The increases in the intensity of attention directed by the driver to the object of interest in the focal or central visual field result in an increase in response time and errors in reacting to stimuli in the visual periphery. The effect increases the greater the eccentricity of the stimuli. The effect of UFOV on driving is best illustrated in a study by Bian et al. (2010). The foveal task for the drivers was to maintain a specific following distance to a lead vehicle. To vary the difficulty of this foveal task, the lead vehicle varied in speed. The peripheral visual task consisted of changes in a light array located above the roadway. The eccentricities of these lights varied from about 3 to 8.5 deg. Driver performance in detecting these lights reduced with a higher workload in following the lead vehicle and response time to the lights increased with their eccentricity.

The UFOV has become an increasingly common means of measuring driver attention, especially for the older driver (see Chapter 6). A meta-analysis of studies examining the relationship between UFOV and driving performance has been conducted by Clay, Wadley, Edwards et al. (2005). The study found a very strong relationship between performance on a UFOV test and driving outcomes. Poorer driver performance on this test was reliably associated with poorer driving performance.

The UFOV research study reinforces the view that driver attention, both its capacity and limitations, is central to understanding driving behavior in a complex environment. Note that while UFOV is defined within the context of visual attention, it should be considered an indicator of the driver's general ability to attend to and process information in the road environment. Tests assessing drivers' UFOV have been suggested as part of the licensure process.

TASK DEMAND AND RESOURCE ALLOCATION

Task demand is a part of the attention process. Attention is the means by which resources are directed to meet the demand required of an individual task. In the involuntary alerting process, attention is captured temporarily with resources initially devoted at a lower, perceptual level. In the case where visual attention is diverted involuntarily, visual alerts elicit an orienting and fixation process from the driver to determine what the nature of the alert is and what further resources, if any, may be required. The driver may choose not to proceed further and shift attention to another task or may decide that further processing of the event is required. In the initiation of control action like braking, the stimulus event such as an abrupt-onset hazard may require emergency braking. The braking action for an experienced driver is automatic in that further information processing is not required nor is any attention required to execute the braking response sequence of foot movements.

In steering control, visual attention is applied to those roadway elements or markings, such as the tangent point in a curve, that provide information about roadway change in direction. Visual attention to these stimuli is maintained as long as road curvature exists. The stimuli result in continual updates to steering wheel inputs so that changes in curvature will result in changes in steering wheel angle. This is true of many highly learned perceptual-motor sequences that experienced drivers carry out in response to specific stimuli including braking, steering and speed control, as well the shifting response sequence in vehicles with manual transmissions. These control actions are initiated by an attentional mechanism that focuses on a specific stimulus or set of stimuli that trigger the corresponding action sequence. Notably, this automated action sequence is highly economical with regard to driver resources, particularly cognitive resources. The automation of these actions allows attentional and other cognitive resources to be allocated to other tasks. For this reason, as well as for the reason of efficiency in performance, automating these motor skills through sustained practice is necessary for driver proficiency.

Apart from these automated skills, however, drivers need to prioritize driving tasks by voluntarily focusing attention on those tasks not involving automated skills. Most of the driving behavior is composed of tasks that require controlled behavior, behavior that is not subject to automation through sustained practice. Such behaviors are initiated by the driver voluntarily attending to the stimuli needed to execute the task. As the task is controlled, not automated, the driver needs to maintain the focus of attention as long as is needed to process the information supporting achievement of the task goal. This does not mean that controlled tasks do not include perceptual-motor components, but rather that these components are not automated and need to be consciously controlled. For example, the vehicle-following maneuver necessitates the use of accelerating or decelerating actions to maintain a safe following distance. The particular control actions involved can be executed without the driver attending to the details of foot movements, but the visual attention to distance maintenance and the initiation and secession of accelerator or brake movements are controlled. The controlling of these components requires the driver to expend both the resources of visual attention and the perceptual processing involved in distance judgments. Unlike the automated braking response to rapidly expanding image size in collision avoidance, the process of adjusting and re-adjusting distance in the vehicle following depends on higher-level cognitive functions, including those of distance perception and focused visual attention even though the low-level motor actions of accelerator and brake pedal use may be the same.

Each of the maneuvers described above involves the expenditure of driver resources which are inherently limited. Thus, each maneuver when executed involves the expenditure of resources. The longer and more complex the maneuver, the more of the driver's resources will be expended. For driving maneuvers, attending to and controlling actions in a maneuver requires that attention be continually focused on the task at hand. Each maneuver has a specific task goal that is kept in mind by the driver during the maneuver execution. For example, the overtaking or passing maneuver has the task goal of overtaking the lead vehicle and returning the driver's vehicle to the starting lane of the maneuver in front of the lead vehicle. The task goal, as well as the rest of the task components such as following distance, is held in working memory as the task unfolds.

TASK PRIORITIZATION

As many driver tasks and maneuvers require the expenditure of limited resources, the driver must prioritize the performance of some driving tasks over others. Ideally, these driving tasks will always have priority over all non-driving, i.e., distracting tasks. The prioritization of tasks by drivers is *value-driven* and is developed through driver training and experience. Value-driven means that the driver attends to those tasks, whether driving-related or distracting, that are deemed to have a higher value than others. The relative value to the driver is that the task achieves a goal for the driver that is valued more than others. A driver may decide to overtake another vehicle because the overtaking task is deemed to achieve a goal (e.g., arriving at the destination earlier) than the existing task of following the lead vehicle. Part of the value-driven prioritization scheme must include the perceived risk associated with the task, either its execution or the failure to execute it. As noted previously (Chapter 4), hazard perception task execution is partly determined by the risk assessment associated with perceiving potential hazards. Risk enters into the prioritization scheme of a driver because experienced drivers recognize the inherent risk in driving. Achieving driving task goals, such as those goals that are part of any maneuver, must be balanced against the risk associated with that achievement. Risk assessment is therefore a modulating factor in determining the importance of task goals including the ultimate goal of arriving at the destination safely. Risk-averse drivers will prioritize those tasks which avoid exposure to hazards, such as collision or loss of control, even if this results in greater time and effort in reaching the ultimate destination. Driving more slowly, avoiding heavy traffic conditions, employing greater following distances, and not driving at night or in inclement weather are among those behaviors characteristic of risk-averse prioritization schemes. The reverse holds for the risk-seeking driver. Exceeding the speed limit, particularly under high-risk conditions such as curve driving, intolerance of following others and a resultant preference for risky overtaking maneuvers, and a tendency for last-moment reactions to hazards are all characteristics of a value-driven scheme that prioritizes exposure to risk as part of the process of reaching a destination in minimal driving time.

Most experienced drivers fall into the risk-tolerant category where some level of risks is acceptable but that deliberate seeking of driving risk is not. The need to arrive safely without incidents is given higher priority than arriving at the destination as quickly as possible. Risk-tolerant drivers avoid unnecessary risk, but not to the point where risk aversion interferes with a reasonably timely arrival at the destination.

Prioritization is largely based on decision-making processes of the driver, which results in a shift of the focus of attention from one task to another according to which task is deemed more important. These processes are a product of a *naturalistic decision-making* process (Klein, 2008). The emphasis of this process is not the deliberative process commonly ascribed to decision-making in other environments but a process that is based on experience and training focused on the development of pre-determined responses to prescribed situations. These responses to particular events are termed *recognition-primed*, reflecting the fact that driver response is triggered by the driver's recognition of a particular event as requiring a specific response or perception-action sequence of responses. This driving behavior may be described as relying on preformed schema or response patterns stored in the driver's long-term memory.

This ideal picture of the competent driver who completes the strategic task of driving to a desired destination is, unfortunately, not always the case. The value-driven prioritization scheme may produce quite different results for other drivers. A driver lacking training and experience who has a value-driven response system that does not prioritize the driving tasks appropriately or fails to prioritize driving tasks above distractions will experience very different outcomes than the idealized version. While different task prioritization schemes can vary greatly among drivers, the most dramatic driver failures occur when secondary tasks, either driving-related or not, introduce demands on driver resources that result in reductions in driver overall performance. While the most publicized of these tasks, like texting, are often associated with fatal collisions or loss of control, many other less publicized secondary tasks have resulted in less catastrophic but increasingly frequent driving accidents and incidents.

SITUATION AWARENESS

Driver attentiveness controlled by a value-driven prioritization scheme should ultimately generate an awareness of the situation both within and without the vehicle. The term *situation awareness* (SA) was originally defined in reference to fighter pilot combat awareness of a tactical situation (Endsley, 1995). The fighter pilot must maintain an awareness or conscious knowledge of all elements that might affect the pilot's mission including awareness of emerging threats, the position of the pilot's aircraft and other friendly aircraft, and the current state of the fuel and weapons state of the aircraft. The mission-critical knowledge of the situation must be continually updated. This requires the pilot to maintain and refresh most of these elements of the situation in working memory, a limited capacity resource.

A more recent expanded interpretation has been developed for application to driving (Gugerty, 2011). This view considers SA as the real-time driver knowledge of the current driving situation in what is considered a dynamic model of the ongoing situation. The model is updated with perceptual and cognitive processes including automatic, pre-attentive processes as well as recognition-primed decision processes and conscious-controlled processes. The first two of these processes place limited or no demand on cognitive resources, but the conscious-controlled processes place heavy demands on these same resources. While Gugerty agrees that automated vehicle steering and speed control are automated processes that place no demand on cognitive resources, the ambient vision processes involved in attention capture are important in maintaining situation awareness. Abrupt-onset hazards such as objects entering the roadway may be detected from ambient vision and capture focal attention. With attention capture, ambient vision may impact the resource demands of SA even though they do not do so directly. Recognition-primed decision processes are brief, typically less than 1 sec, responses involved in events such as changes to traffic lights, lane change maneuvers and the like that involve only brief demands on attention and cognition. Conscious-controlled processes, by contrast, are involved in tasks such as navigation in unfamiliar routes, detection and categorization of potential road hazards, and related events. These tasks may involve more extensive cognitive processing involving spatial relationships, behavioral inferences about driver and pedestrian behavior and storage, and the continuous updating of position and prioritization of these events in working memory.

A general state of heightened awareness when driving is a key ingredient to the maintenance of SA. This heightened state of awareness increases the likelihood of detecting collision and loss of control threats; the proximity of other traffic and hazards; awareness of traffic regulatory mechanisms such as traffic lights and signage; awareness of current speed, stopping distance, and lane position; and other driving-related tasks. The heightened alertness needed to maintain SA requires considerable effort on the part of the driver. Maintaining SA for prolonged periods of time at this level of alertness is difficult for many drivers, particularly older drivers and those drivers who are fatigued or have impaired health.

Research conducted on driver SA requires a focus on different measures than typically used for other research issues. As SA is fundamentally an issue of knowledge of driving-related events stored in working memory so a probe or query of this knowledge is one of the best means of determining SA. A comparison of SA among young (aged 16–25 years), middle-aged (40–50 years), and older (65–80 years) drivers was conducted on their ability to attend and store important information from the driving environment (Bolstad, 2001). Older drivers had lower SA than younger drivers. Young drivers did not differ from middle-aged drivers in their SA.

A number of factors may contribute to older drivers' lower SA including reduced working memory and visual attention capabilities (see Chapter 6). A later study of the effects of age on SA compared a much younger group of drivers (Kass, 2007). A young, novice group (14–16 years) was compared with a young, experienced group (21–52 years) using the query method and measuring the number of infractions committed such as speeding, road edge crossings, pedestrian collisions, and others in a driving simulator. Novice drivers committed more infractions and had lower SA than the experienced driver group. The performance of both groups was negatively affected by cell phone use.

The lower SA of young, novice drivers can be attributed to a much lower visual scanning activity for novice drivers compared to experienced drivers. A driver cannot attend to and store in working memory information unless it is detected and that depends almost exclusively on visual scanning and search. The second element is likely to be the tendency of teen drivers to have much higher risk thresholds than other drivers. Teen drivers who do detect potential hazards may not classify them as such due to these higher thresholds.

In both of these studies of SA, limits to SA were likely the result of limited cognitive resources (working memory) in one case and limited use of resources (visual scanning) and risk threshold levels in another. These factors may also affect the value-driven prioritization scheme mentioned above that is used to assign the most important hazards which have priority for driver resources. Perceived risk of a potential hazard is likely to differ greatly between a teenage driver and an older driver. This risk threshold difference will affect the attention devoted to potential hazards as well as other driving issues. However, the data show that when cell phone use is available to younger drivers the negative impact on SA is the same. The value of attending to the cell phone was more important than addressing driving-related issues.

DUAL-TASK INTERFERENCE

At any moment within the overall driving task domain, drivers will likely expend resources on a single primary task. The primary driving task is generally, but not always, a task related to the control of vehicle direction and speed, collision avoidance, hazard perception, or navigation. Each primary task has associated subtasks which are vulnerable to interference from secondary tasks. The secondary tasks may or may not be driving-related. In the case of driving-related tasks, a secondary task may demand resources used in the primary driving task. If the demand is great enough, performance on either or both the primary and secondary tasks may decline. Thus, the determining factor of secondary task interference is what driver resource is impacted and whether this resource is currently in use by a primary driving-related task.

For decades, researchers have investigated the role of dual-task performance in a variety of task domains. However, many of these were very simple tasks conducted in controlled laboratory environments. Their relevance to driving is often relatively remote and, as a result, difficult to generalize to actual driver performance. Nonetheless, some studies have investigated dual-task interference in driving that reveal the degree to which physical, perceptual, and cognitive resources are demanded by individual driving tasks.

BIOMECHANICAL INTERFERENCE

Physical or biomechanical interference occurs when the secondary task imposes demands on the driver's ability to manipulate controls necessary to support the primary driving task. The most common form of biomechanical interference occurs when one or both of the driver's hands needed for the operation of a control, such as the steering wheel, are engaged with a secondary task. This task might be driving-related such as the operation of headlights or windshield wipers both necessary for the primary task of vehicle control. In this case, the removal of the hand from the steering wheel and returning it is relatively brief. Moreover, the demand to initiate these actions is relatively low allowing the driver to time their execution for periods when the demand from other driving tasks is low. Biomechanical interference of controls is more often due to a task unrelated to driving. For example, one hand may be used to hold or manipulate objects such as a hand-held cell phone, a beverage, or a control on an entertainment system. Unfortunately, these often are not brief engagements with interactions that may last minutes. Even if the driver does nothing but hold on to an object, steering control may be affected. The steering hand may be positioned non-optimally on the steering wheel reducing the extent of steering wheel angle (SWA) displacement. The positioning and limited strength of one hand reduces the amount of torque that can be applied to the wheel. Hand-held phone devices have been shown to produce biomechanical interference with steering control whether in passenger cars or heavy vehicles (Brookhuis, 1991; Ishigami and Klein, 2009; Tijerina et al., 1995). Moreover, single-handed driving may impair the driver's ability to execute emergency maneuvers such as avoiding abrupt-onset hazards entering the roadway, when encountering strong crosswinds, or in skid recovery control.

Biomechanical interference is not limited to hand controls. Interference with foot controls has also been documented such as that which occurs in the case of unattended accelerations due to improper positioning of floor mats (Lee, 2020). More often, impairment in the use of vehicle foot pedals may be caused by the pedals themselves. Interference of the brake pedal has been shown in driver foot movements from accelerator to brake pedal. The case of brake pedal interference has been found to occur at a significant rate. In a study of driver foot movements in startup and parking maneuvers, some 20% of 3300 events involved "brake pedal hooks" where the driver's foot catches the side of the brake pedal when moving from the accelerator pedal to brake pedal (McGehee et al., 2016). The result of hooking the foot on the brake pedal impairs the movement of the foot from the brake pedal. If a driver moves the left foot to the brake pedal, pressure on the brake pedal is impaired by blockage of the right foot. Moreover, pressing harder on the brake pedal will put further downward pressure on the accelerator from the entangled right foot. As this action may be in response to an urgent braking requirement, there may be little or no time available for the driver to correct the problem.

SENSORY-PERCEPTUAL INTERFERENCE

A more common problem of driving task interference than that of biomechanical interference is the result of demands made on the driver's sensory input systems, especially visual. The tasks of vehicle steering control and the control of speed are heavily dependent on driver visual input from the roadway with occasional inputs from in-vehicle instruments such as the speedometer. At the level of sensory function, steering control is primarily dependent on a visual system limited to a central field of about 10 deg in diameter for both eyes combined. Ambient motion perception from the visual periphery aids the driver in the perception of speed and in the control of lane position.

The central visual field is particularly important for steering control in lane changes, negotiating intersection turns, and perception of road curvature. The driver focuses the central field on where the driver intends the vehicle path even before steering wheel input begins. If the driver shifts the focus of the central field to another area prior to or during these maneuvers, the maneuver may fail. Driver vision is fundamentally a sensory system and cannot function effectively without the proper orientation to the roadway where it is needed and when it is needed to support steering control. In some tasks, such as tracking road curvature, steering control updates are occurring at rates of about 1/sec so interference with visual inputs for longer periods can have catastrophic consequences.

Due to its importance in-vehicle control, vehicle control will be affected by any task that shifts the orientation of the visual field away from the roadway visual cue that supports the particular control task. This is true regardless of whether or not the task is driving-related. However, when the driver moves the visual focus to the inside of the vehicle, such as toward a display device, a change in visual accommodation from far to near focus is required. Changes in visual accommodation take time as does re-accommodation when the driver looks back to the roadway. Accommodation response times average 0.34 sec far-near and 0.35 sec near-far (Heron, 1999). These response times increase with age. Subsequent visual search of the in-vehicle device display contents or operating a particular control takes even more time away from the roadway.

The design and positioning of these in-vehicle displays with respect to the driver's on-road line of sight is important in how they affect driver vision. Horizontal eccentricities from the driver's line of sight are one factor known to affect driver behavior. Eccentricities of 35 deg or more have an increasing impact on lane-keeping performance and driver reaction time to events outside of the vehicle (Wittman et al., 2006). Even brief shifts of vision from the road to attend to an in-vehicle display task can affect vehicle control (Tsimhori et al., 2007). The impact on vehicle control of the in-vehicle task increases as the steering control task becomes more demanding of driver resources.

Of those in-vehicle devices that impact vehicle control, including steering control and braking response, hand-held devices like the cell phone have the most impact. Not only do these devices have potential biomechanical interference, but their very purpose and design requires that vision be diverted away from the roadway to the device display and control system when an incoming call is received. Even before answering the call, attention will be diverted from the roadway to determine the identity of the caller and determine whether answering the call is desired. Thus, the driver is compelled to respond even if there is no intention to answer the call. Reviews of cell phone studies show a consistent impact on steering control and braking response with cell phone use (Oviedo-Trespalacios et al., 2016). Among other factors, visual distraction from the roadway is one reason for poorer driver vehicle control behavior. The distractions from in-vehicle displays interrupt the normal perception-action sequence of steering control and delay the detection time of road events that require braking.

Perhaps the most intrusive of in-vehicle devices are those that require text-based interactions where the driver must read and type communications to remote sources. In a review of in-vehicle texting, typing, and reading text messages affected a variety of driving tasks including detection of road events, lane position control, speed control, and following distance (Caird et al., 2008). Among the most affected of the driving behaviors measured were eye movements with reading only slightly less distracting than typing. Hand-held devices had a particularly strong effect on control of lane position.

Hands-free devices have been introduced as an alternative to hand-held devices in an attempt to mitigate the interference of cell phone use with driving tasks. The biomechanical interference with driver controls would be eliminated and hopefully other forms of interference as well. Reviews of studies comparing hand-held and hands-free cell phone interference with driving behavior have found that both types of cell phones interfere though in somewhat different ways (Ishigami and Klein, 2009). Both impaired the detection of road events and increased the braking response time of drivers. Drivers were found to have a compensatory slowdown in speed with hand-held but not hands-free devices. Stopping distance and following distance were greater for hands-free than for hand-held device use.

Attempts have been made to reduce the impact of texting on driver behavior as well.

Speech-based text entry allows the driver to transmit text using a hands-free device. Text entry via speech has been found to impair driving, though not as much as held-held text entry devices (He et al., 2014). Both speech-based and manual text entry affected the control of speed, lane position, and following, though the effects of speech-based entry were somewhat less. Compensatory increases in following distances were again found for the hands-free device.

Cognitive Interference

Shifting focal visual attention is either a response to a strong alerting stimulus, such as a bright light or sound, or an intentional response to direct attention to what the driver perceives to be a more important task. A response to an alerting stimulus does not necessitate a shift to focal visual attention beyond that needed to identify the source and determine its importance relative to the primary driving task. The design of an in-vehicle display system has to consider the impact of an alerting stimulus on driving behavior. For the vehicle designer, the event which necessitated the alerting stimulus must be of sufficient value that distracting the driver from the primary task (e.g., vehicle control) is justified. In-vehicle display and alerting systems that do not design with this justification for attracting driver attention in mind may result in a *design-induced error*, which could result in a collision or loss of control.

Cell phones are an example of how a system designed for mobility became a source of designed-induced driver error as a consequence of its use in moving vehicles. Telephones were not designed originally with vehicle use in mind and were not considered part of the vehicle design process. Rather, it was the mobility of the end-user afforded by cell telecommunications technology that drove its design into widespread use in vehicles. However, basic telephony requires that the recipient be alerted to the incoming call. The distinctive aural alert of the cell phone ring has a demand characteristic beyond those normally found on in-vehicle systems. The cell phone design implicitly assumes that the driver's task priorities include that of telephony regardless of the potential impact on other tasks that the driver may need to perform. The driver must intervene in the telephony process by disabling the alert system to prevent distraction.

Higher-level processes, like the focal visual attention involved in driving, appear to be at the greatest risk of interference from in-vehicle devices than either the sensory-perceptual or biomechanical resources. Moving the central visual field usually means that the focus of visual attention and the cognitive processing that may accompany it shift as well. Drivers may simply scan the in-vehicle displays briefly to capture parameters such as vehicle speed. Scanning primary instrument displays requires minimal visual attention and processing, if any. Absent an alert or warning stimulus, drivers attend to tasks inside the vehicle because they have decided that a more important task needs to be attended to than any outside the vehicle. In the case of cell phone use, the distinct ring elicits attention to a task that drivers often cannot ignore even though they may not wish to talk to the caller.

The second stage of cell phone telephony is a decision to answer or not. This requires visual attention to the device to determine caller identification, followed by a response to either complete the connection or ignore and send the caller to voice mail.

Part of this decision should be influenced by the existing driver task environment, especially driver task loading. However, the driver may decide that receipt of the call is too important to miss and answer the call.

This is a part of the *prioritization* of tasks mentioned previously that is an important element in driving. This necessarily involves not only an assessment of the driving task demand but the risk associated with making the call a priority over driving tasks.

The first two stages of cell phone telephony are engaging higher-level cognitive processes associated with decision-making and risk assessment. This is true whether the cell phone is hand-held or hands-free. As cognitive processes, like decision-making, are serial in nature, responses to decisions regarding driver events such as collision avoidance will be slowed even before a conversation has begun. Once the driver has decided to answer the call, the heaviest impact on driver resources begins. These resources are those cognitive processes devoted to speech production and speech comprehension.

In the case of aural cell conversations, the conventional means by which cell phones are used, the driver does not need to focus continual visual attention on the device display. Instead, the driver may re-focus attention on the roadway. The driver may believe that the risk associated with cell phone use has thus been effectively eliminated. However, the process of conversation requires the process of continuous attention to the caller's speech in order to follow the contents of the conversation and its meaning. Beyond the required focus of aural attention to the conversation, the driver must use both working memory and higher-level speech processing in order to understand what is said. When the driver has the turn in the conversation, speech generation processes begin. These are even more resource intensive as speaking requires the retrieval of vocabulary and grammatical rules in order to make the speech comprehensible. It is not surprising that research has revealed that high levels of *inattention-blindness* may occur during cell phone conversations (Strayer and Drews, 2007). In the case of inattention-blindness, the driver may "see" the driving event but fails to "perceive" the event because cognitive processes are preoccupied with conversational processes. In this study, driver processing of less than 50% of information in the driving environment occurred during cell conversations. The study also found that normal passenger conversations were found to be less intrusive than cell conversations possibly due to the ability of the driver to better balance the passenger conversation with the driving task. Other studies, however, have found that both types of conversation produce similar levels of interference with the driving task (Horrey and Wickens, 2006). It is likely that the degree of interference from passenger conversations will depend on the perceived importance of the conversation contents to the driver as well as the ability of the driver to prioritize the driving task (Drews et al., 2008). Conversational activities, whether conducted remotely with a cell phone or with a passenger, carry a substantial risk of interference with essential driving tasks.

Thus far, the analysis of a cell phone conversation has been conducted with the assumption that it is normal aural communication. The protocols of these conversations are the same format or structure whether carried out on a cell phone or any other type of phone. Cell phone conversations conducted via texting, however, necessarily require a high degree of focused visual attention to the display. Unlike aural communications, text-based communications do not allow the driver the freedom in communications such as interrupting the conversational flow to synchronize the text with the driving task. Pacing the conversation is driven by reading and entering text which is, relative to aural communications, time consuming as well as requiring sustained visual attention. For standard cell phone use, the participants compose, send, and reply to text messages. This requires the same fundamental processes as aural communication with the additional task of reading and typing text. Not surprisingly text-based cell communications will likely consume arguably more driver resources than any other driving or distracting task whether inside or outside the vehicle. A meta-analysis of 33 studies of the effects of texting on driving shows the widespread impact of texting, both reading and typing, on driving tasks (Caird et al., 2008). The texting task affects driver visual scanning and search, detection of road events, collision detection, lane position control, speed control, and the control of following distance. Of the driving tasks measured, driver eye movements were the most affected by texting. This would be expected of a task like texting which demands such high levels of visual attention. A later study found that texting also increased the driver's response time to the brake lights of a leading vehicle as well as affecting vehicle control (Drews et al., 2009).

A recurrent theme throughout the issue of multitasking in driving, whether the tasks are driving-related or are simply distractions, is the issue of *selective attention*. Visual attention is a form of selective attention where the focus of attention is directed to a visual stimulus or stimulus array. The visual stimulus could be something as simple as a bright, flashing emergency beacon that captures the driver's attention or it may be as complex as an electronic road sign. The voluntary, selective attention to a stimulus, such as an in-vehicle display or a cell phone display, is a part of the driver's prioritization scheme or plan. Thus, the driver attends to stimuli, other than those that elicit an involuntary attentiveness, to support a particular task that the driver considers important at that particular time. This does not necessarily mean that the driver's selection of the task is correct for the circumstance but rather that it supports the driver's priorities. The driver is attending in this case to a value-driven action which supersedes other less valued activities. This applies to cell phone tasks and other in-vehicle display systems unrelated to primary driving tasks. Visual focal attention means that the driver's visual system is now focused on that stimulus and only that stimulus.[1] It also means that the driver's visual system, from visual receptors to higher-level image processing, is engaged in the stimulus. Tasks that involve the physical movement of a control such as braking cannot share this motor action with any other control. However, automatization of the braking tasks does mean that little or no selective attention to the task is required of the driver during its execution.

DRIVER TASK – RESOURCE RELATIONSHIP

The complex relationship that exists between tasks, either driving or non-driving, and the driver resources needed to carry out these tasks can be illustrated by cross-referencing the resources, physical and mental, with their respective tasks. In Table 5.1, commonly used driver resources are identified for selected driver tasks ranging from basic steering and speed control to the distractions of generic, non-driving cognitive activities. While demand levels of these tasks on driver resources can vary depending on the specifics of the task and prevailing task environment, some tasks will always place some demand on specific driver resources. The presence of a task-resource demand for these tasks is identified with an "X" in Table 5.1.

A task-resource matrix presented in Table 5.1 needs to be constructed for complex task environments such as driving to ensure that new tasks such as those required of new in-vehicle technologies are integrated into the driving system with minimal impact on driver resources. Table 5.1 should be considered an example of how this might be done but, in this case, is simplified for the purposes of illustration.

DRIVER TASKS

The selected driver tasks in Table 5.1 need to be defined in more detail as only a limited space is available in the table. The first task listed, steering control, refers to the basic perceptual-motor task involved in maintaining directional control and lane position in straight and curved roadways and intersections. Speed control refers to the basic perceptual-motor task of maintaining vehicle speed at a desired level and accelerating or decelerating when needed. Collision detection refers to driver detection of an abrupt-onset collision hazard, which may occur in the driver's visual periphery or in front of the driver's vehicle. The collision avoidance response is the driver's response to an abrupt-onset hazard. It typically consists of a combination of braking and maneuvering. The perception of potential or slow-onset hazards refers to near-future collision hazards or loss of control hazards. Wayfinding refers to the conventional means of developing and using cognitive maps in navigating familiar routes such as those that might

TABLE 5.1

Driver Resource Demand for Selected Driving Tasks

Driver Task	Ambient Vision	Focal Vision	Ambient Audition	Focal Audition	Sensorimotor (Pedal)	Sensorimotor (Wheel)	Sensorimotor (Hand Control)	Cognitive Processing (Working Memory)	Cognitive Processing (Language)
Speed Control	X		X		X				
Steering Control	X	X							
Collision Detection	X	X							
Collision Avoidance Response					X	X			
Hazard Perception (Slow-Onset)		X						X	
Wayfinding (Cognitive Map)		X						X	
Navigation System		X		X					H
Vehicle Instrument Display		X					X	X	
Advanced Driver Assistance System		X		X					
Radio/CD System		X	X						

(*Continued*)

TABLE 5.1 (CONTINUED)
Driver Resource Demand for Selected Driving Tasks

Driver Task	Ambient Vision	Focal Vision	Ambient Audition	Focal Audition	Sensorimotor (Pedal)	Sensorimotor (Wheel)	Sensorimotor (Hand Control)	Cognitive Processing (Working Memory)	Cognitive Processing (Language)
Passenger–Driver Communication				X				X	X
Hand-held Phone		X		X			X	X	X
Hands-free Phone				X				X	X
Hand-Held Phone Texting		X					X		X
Road Signage		X							X
Cognitive Activity (Not Driving-Related)								X	X

be used in daily commutes. The operation of navigation systems, however, refers to driver navigation using in-vehicle navigation display systems that employ aural communication for driver alerts and route directives. Vehicle instrument display systems are the primary means by which the driver is informed as to the current vehicle state such as speed, fuel, headlight status, turn signal, engine, and other status displays and include the use of associated controls. Advanced driver assistance systems (ADAS) represent a class of driver assistance systems intended to improve driving safety. They include advanced collision avoidance systems (ACAS), which detect a collision hazard and apply braking automatically; lane deviation warning (LDW) systems, which alert the driver when the vehicle is drifting out of the lane; and a variety of parking assist and drowsiness detection systems, which typically include audio and visual alerts and may include automated vehicle control inputs. Radio/CD systems are standard in many vehicles and provide passive news and entertainment programs aurally with minimal driver inputs. Passenger–driver communications refer to routine communications with passengers regardless of whether the topic is driving-related or not. The hand–telephone task refers to cell phones and similar devices that are held in one hand during aural communication and includes normal tasks involved in the use of the device. The hands-free phone refers to devices which allow the driver to transmit and receive voice communications through in-vehicle systems which both have a visual display and allow driver speech input. Hand-held texting devices are those that allow text-based transmission and text display reception with the device held in one hand. Road signage refers to signage-containing text which has to be processed at a level sufficient for the driver to understand. The last task refers to general driver cognitive activities unrelated to driving.

DRIVER RESOURCES

Pre-attentive ambient vision and the controlled process of focal vision are among the most important driver resources, with the latter resource in greatest demand in driving. *Ambient audition* refers to undifferentiated background noise such as from the vehicle engine or from passengers not communicating among themselves. *Focal audition* refers to focused listening as in alerts and warnings or to communications considered of high value to the driver. Sensorimotor resources refer to low-level, automatized perceptual-action sequences for pedal, steering wheel, and hand control operations. The next driver resource in Table 5.1 is the high-level cognitive processing involved in working memory storage and retrieval. The last resource is the cognitive processing involved in language understanding and product, which includes language in road signage, passenger communications, and written text. Generally, language processing involves working memory in passenger communications as well in order for the driver to keep track of the meaning of the last utterance or expression from the passenger.

SUMMARY

Task demands on driver resources are explored in this chapter with particular emphasis on driving as a multitasking skill executed in a complex, dynamic environment. In such an environment, drivers are compelled to distribute tasks over time. How these tasks are distributed is determined by the driver's value-driven, prioritization scheme. This scheme allows the driver to attend to different tasks that individual task-resource competition is kept to a minimum. Competition among tasks includes biomechanical or physical, sensory-perceptual, and cognitive. While some tasks at the lower sensorimotor level, such as braking, can be automatized through repetition, others cannot. These tasks typically involve focal attention and higher cognitive processes which are serial in nature and are in competition for a single resource. Studies examining interference among driving tasks due to this competition are reviewed. These studies reveal the high cost to performance when driver resources, particularly cognitive resources, compete. A matrix of driver tasks and driver resources is developed to illustrate the competition and to reveal the costs of non-driving tasks on driving behavior.

NOTE

1. The idea of *divided* attention in complex task environments like driving is unlikely if the assumption of a serial central processor is accepted. Thus while shifting attention between tasks is possible, dividing attention to support more than one cognitive task is not possible in a serial cognitive processing system.

6 Aging and Driving Behavior

In the U.S., drivers over the age of 65 now account for nearly 20% of licensed drivers. This number will increase by 25% by the year 2030 or about 70 million licensed drivers. Some of these older drivers may not be capable of operating their vehicles safely in all conditions. An analysis of the changes in sensory, perceptual, motor, and cognitive functions that occur as a result of normal, healthy aging aids in understanding driver behavior when the human operator is no longer functioning at peak efficiency. This impairment in performance may, in turn, increase the likelihood of an accident under some circumstances.

An understanding of the effects of aging on driver performance requires an analysis of all of the physical and mental processes involved in the driving task. Only in this way is it possible to determine the impact of aging on driving performance, which, in turn, will allow for more targeted and effective interventions. The analysis of the effects of normal, healthy aging on driving performance also helps isolate critical physical and mental processes required for competent driving performance.

The literature on the effects of aging and driving generally includes some studies which define aging as beginning at the age of 60, with the majority of studies using an average of 65 or older as defining the older age group. It should also be noted that the specific age at which the term "older driver" is defined has changed over the decades for a variety of reasons including the acceptance of the fact that individuals can reach older ages and remain active drivers even at 85 years or more. However, individuals beyond the age of 75, defined here as *elderly* drivers, are often difficult to recruit in large numbers for experimental purposes, so a definitive analysis of their performance level has been more difficult to achieve.

In order to isolate physical and mental processes impacted by aging, the first section of this chapter will address aging effects on basic visual and non-visual sensory and perceptual systems relevant to the driving task. The aim of this section is to identify the average and the variability in sensory function for given age groups: younger age drivers between the age range of 20 and 59 years and older drivers at and beyond age 60. Where data are available, age groupings above the age of 75, elderly drivers, will be identified as a separate group.

The second section describes the effect of aging on driving behaviors in a wide variety of tasks. While some data will be provided on non-driving laboratory tasks, the emphasis in this section will be on performance

DOI: 10.1201/9781003454373-6

in driving environments such as driving simulators or real-world field studies in actual vehicles. This section will include studies where older drivers have been defined as 60 or above although most studies use an age of 65 or above. It will address the performance of older drivers' physical and mental processes as those processes manifest themselves in more complex tasks. These include simpler perceptual-motor tasks such as those involved in the turn or emergency braking. They also include tasks involving higher-level, cognitive components such as multitasking and situation awareness.

The third section of this chapter will address the problem of using accident statistics in an attempt to isolate specific demographic characteristics such as aging as proximate causes of driving accidents. Specifically addressed will be the misinterpretation of older driver accident data with emphasis on the *low mileage bias*, which occurs when accident data are defined without reference to annual miles driven per age group.

VISUAL AND NON-VISUAL SENSORY SYSTEMS

A more detailed description of sensory, perceptual, and cognitive functions is described in Chapter 1. This section addresses the impact of aging on those sensory-perceptual processes that have relevance to driving behavior. However, a discussion of these processes does not necessarily mean that a decline results in unsafe or even substandard performance.

Visual function is the most important sensory system in driving. Various components of vision show a decline with age, but they decline at different rates. Furthermore, and most importantly, individual sensory-perceptual components have differing levels of impact on driving behavior.

VISUAL ACUITY

Visual acuity (VA) is the measure of vision used most commonly by regulatory authorities in driving and is usually the test the older drivers will need to pass for license renewal. Snellen acuity levels of 20/40 (6/12 m) is a common minimal VA required for driving though lower levels such as 20/60 (6/18 m) may be used in some areas during daylight hours. Note that this test is a measure only of central vision (1 to 2 deg of the central visual field) and is generally measured only under photopic conditions.

Aging effects on VA have received much attention from the vision research community in studies of driving behavior although it is only one measure of visual function. The sample sizes are relatively small and do not differentiate age groups at a level of detail that allows comparisons among age groups, particularly those over the age of 65. Large-scale studies which allow the comparison of the effects of aging across age groups are relatively rare. These year-by-year data are important to identify the mean or median VA per age group as well as the variability of VA scores for each age group.

The seminal study by Haegerstrom-Portnoy et al. (1999) tested a sample of 900 older adults ranging in age from 58 to 102 years of age is one such example of large-scale studies allowing comparisons among individual age groups. The study provides the opportunity to compare average and individual variations in VA reported in each age group. Standard measuring techniques of visual function were used. Individuals with ocular disease were not excluded from the sample. This study thus provides a sample more representative of older individuals in the driver population where individuals will continue to drive even with eye diseases. While photopic VA with high-contrast targets declined systematically with age, VA of 20/40 or better was maintained with habitual correction until age 90. Variability in VA increased continuously beginning at age 85. When photopic luminance was reduced to 15 cd/m² and contrast to 11%, the effects of aging on VA were more pronounced. VA dropped below 20/40 even for the youngest subjects declining to 20/125 (6/38 m) for those aged 85 years. For those aged 90 years or older, VA dropped below 20/200 (20/60 m). Variability in VA among individuals in the study sample was generally higher in the low contrast and low luminance condition than in the high contrast, high luminance condition. Clearly, older-aged individuals are more likely to be affected by reduced levels of luminance and image contrast than younger ones.

VISUAL ACUITY AND DRIVING

The implicit assumption in the widespread use of VA testing normally conducted under photopic conditions is that photopic VA is essential to a predominantly visual task such as driving. As aging results in VA decline, it would be expected that older drivers would have difficulty in some visual tasks. Indeed, the large-scale survey by Kline et al. (1992) found that older-aged drivers have difficulty with a wide variety of visual driving tasks such as detecting dimly illuminated objects and objects near to them or moving rapidly or that are part of a complex array of stimuli. However, no data were provided relating photopic VA to these problems. A study of road sign discrimination distance did not relate to the VA of drivers either older or younger (Evans and Ginzburg, 1985). A more recent field study by Woods and Owens (2005) found that photopic VA did not predict the behavior of older drivers in recognizing road signs, low-contrast road obstacles, or pedestrians in either day or night conditions. Finally, reviews of accident data have been equivocal with respect to photopic VA and driving accidents (Shinar and Schieber, 1991; Woods, 2002).

Photopic Contrast Sensitivity

Unlike VA, which is a measure of spatial resolution, contrast sensitivity is a measure of the sensitivity of the eye to differences in luminance between an object and its background. In Haegerstrom-Portnoy et al.'s (1999) study and in another large-scale study by Rubin et al. (1997), contrast sensitivity was found to begin to decline at age 65 and continued to decline thereafter, reaching moderate loss levels at age 80 and continuing to decline to severe loss levels at age 95. A high correlation (r = .86) was found between VA at high luminance and high luminance contrast sensitivity in the Haegerstrom-Portnoy et al. (1999) study. Photopic contrast sensitivity appears to be a good predictor of loss in visual performance even when VA is controlled (Schieber, 2006). For example, older drivers with low photopic contrast sensitivity have been found to have difficulty reading road signs under photopic conditions (Evans and Ginzburg, 1985).

Mesopic Contrast Sensitivity

Mesopic vision occurs within luminance levels of 0.01 and 3.0 cd/m². This is typical of luminance levels encountered in night driving. Mesopic, unlike photopic, vision involves both rod and cone photoreceptors and their interactions. The interaction between these photoreceptors makes it more difficult to determine the influence of factors that normally affect these two groups when measured separately. Color vision, for example, shifts in wavelength to a shorter length in mesopic vision (a phenomenon called Purkinje Shift).

VA and contrast sensitivity are both affected by low luminance and both decline with luminance levels and aging. However, as these luminance levels approach the upper limits of mesopic vision, as happens in night driving, the photopic measures of acuity which are based on cone photoreceptors, no longer apply. Thus, it is not surprising that photopic VA is a poor predictor of night driving ability (Gruber et al., 2013). Mesopic contrast sensitivity, which declines with age, shows strong predictive values in assessments of the driving habits of older drivers (Puell et al., 2004). The study by Woods and Owens (2005) revealed that the contrast sensitivity of older drivers, unlike photopic VA, resulted in poor object recognition in low luminance conditions. Mesopic sensitivity was also found to be predictive of driving accidents among older drivers in a recent large-scale, prospective study (Owsley et al., 2020).

CONTRAST SENSITIVITY AND DRIVING

Given the decline in contrast sensitivity in older drivers and its potential to increase the likelihood of accidents, particularly at night, a more detailed analysis of the role of contrast sensitivity and night driving is warranted.

Three driving tasks are especially affected by contrast sensitivity problems: legibility of road signage, control of the vehicle, and hazard recognition.

Legibility Distance

One of the driving tasks most sensitive to vision problems is the legibility distance of road signs. Legibility distance is the distance from the road sign at which the driver can read the sign contents. In general, older drivers need to be much closer to signage in order to read them than younger drivers. The problem has been found to increase with age (Owens et al., 2007). In a field study of night driving, older drivers averaged 65% of the legibility distance of younger drivers (Chrysler et al., 1996). Not surprisingly, the problem of legibility distance is aggravated by low road luminance at night (Easa et al., 2010). The effect of road luminance on legibility distance was eliminated in this latter study when luminance was increased from 0.6 to 2.5 cd/m².

Legibility distance reductions in older drivers can have the consequence of not allowing the driver sufficient time to exceed proper maneuvers. The study by Kline et al. (1992) revealed that older drivers maintain excessive speed when entering curves. Reduced legibility distance also affects the time to slow down when entering an intersection or being able to stop before entering an intersection. It may also affect a variety of wayfinding behavior by making street signs nearly impossible to read in order to make route-correct selections. As legibility distance declines for older drivers, it effectively compounds the problems related to slowed decision-making processes common in older age drivers.

Vehicle Control

A potentially more serious problem may arise from reduced contrast sensitivity in older drivers, and that is the problem of maintaining lane position in night driving. Lane position cues are derived from reflective lane markings demarcating each roadway lane. Road edge markings, when available, separate the usable portion of the road from the road edge or from the portion of the road reserved for emergency use. As discussed earlier (Chapter 2), road markings are used to provide continuous steering guidance to the driver in order to maintain lane position. At night, the headlights of the vehicle reflect differentially the markings from the road surface. The higher contrast between the markings and the road surface, the easier it is for the driver to maintain lane position.

In a driving simulator study, older drivers were found to have progressively poorer steering accuracy with low luminance while young drivers' steering accuracy was not affected (Owens and Tyrrell, 1999). However, in the field study of Owens et al. (2020), the behavior of older drivers steering was different. In this study, older drivers hugged the road edge line and reduced speed when luminance was reduced. The older drivers appear to have compensated for the poor visual conditions in these more hazardous, real-life conditions by adopting driving behavior which increased their safety margin by slowing rather than maintaining speed and risking the consequences of poorer steering performance.

HAZARD RECOGNITION

Perhaps even more concerning than either sign reading or vehicle control is the impact of age on the ability of drivers to recognize and respond to road hazards under night driving conditions. One of the features of visual contrast sensitivity is to not only separate objects from their backgrounds but to discern specific features of those objects in order to correctly classify them. In the studies described above, the low luminance contrast sensitivity of older drivers made the recognition of low luminance objects more difficult than for younger drivers. Early recognition of hazards is obviously essential to allow time for collision avoidance behavior such as braking or maneuvering. For older individuals whose decision processing time may be longer than younger drivers, this delayed recognition is problematic. A second or two of delayed recognition of a hazard at a speed of only 30 mph translates to a distance of between 44 and 88 ft of travel.

A second problem arising from the reduction in contrast sensitivity of older drivers is the detection of object features under low luminance conditions. Object features include attributes of an object which help to classify it as a potential hazard. The most important of these are pedestrians who often wear clothing with low reflectance which hides characteristics which might identify them as pedestrians rather than inanimate objects. An object classified by the driver as inanimate and facing the roadway or road crossing, for example, is not likely to be classified as a potential hazard, but an object classified as animate such as a pedestrian is likely to be perceived as a potential hazard. This is determined by extracting visual features which are peculiar to a pedestrian. Owens et al. (2020) enhanced the pedestrian with biomotion features by attaching reflective markings to arms and legs. These markings are known to improve pedestrian recognition in night driving. Pedestrian recognition was enhanced for older drivers in this study when compared to younger drivers.

STEREOACUITY

The provision of two eyes separated horizontally in the head provides the human species the ability to see objects in stereovision, or, more accurately, it provides for stereopsis. The depth perception resulting from stereoacuity or *stereopsis* is, however, in addition to other depth cues such as size–distance and texture-density that may be available. The latter typically play a larger role at distances beyond those of stereoacuity. The importance of stereoacuity is predominantly at shorter distances in the range of 1 to 2 m from the eye decreasing in effectiveness linearly to a distance of about 6 m (Cutter and Vishton, 1995).

Stereopsis declines significantly with age. By age 80, stereopsis is only about one-sixth of that measured at age 7 (Lee and Koo, 2005). The largest rate of decline begins at age 60 when its impact on driving is most likely to be felt. Due to the range limitations of stereopsis, however, only the perception of objects within the vehicle or in close proximity to the vehicle is likely to be affected.

Thus far, studies of older drivers have not found that losses in stereoacuity result in driving problems (Rubin et al., 1994). A study comparing depth judgments of younger and older drivers using 2-D and 3-D displays of video clips found comparable performance (Norman et al., 2008). However, these studies did not focus on tasks performed in the vehicle such as the operation of hand controls requiring more precise depth perception. Additionally, studies on the impact of stereopsis on tasks involve accurate proximity judgments such as those involved in parallel and diagonal parking.

VISUAL FIELD LOSS

Loss of the visual field occurs with aging due to a variety of factors. Declines are particularly pronounced after age 60. Haegstrom-Portnoy et al.'s (1999) study showed visual field loss is largest at the visual periphery (> 40 deg off the foveal axis). The surveys of Keltner and Johnson (1980, 1983) revealed field loss of between 13% and 14% for those over age 65. Those over 65 were found to have four to five times greater incidence of visual field defects than younger drivers. Correlations with accident rates have been found for field loss in both eyes but not in one.

The impact of visual field loss on driving would likely affect the perception of the driver's own vehicle speed since much of self-motion perception depends on optic flow patterns which are strongest in the visual periphery. Furthermore, loss of vision in the visual periphery also increases the likelihood that hazards entering the roadway are less likely to be detected. The study of Bowers et al. (2005) showed that driving performance is affected by visual field loss for maintaining speed, lane position, and steering through curves though aging as such was not a variable in the study.

STRAY LIGHT AND GLARE

During both photopic and mesopic driving conditions, older drivers are more vulnerable to the effects of stray light and glare. This is due to the deterioration of the lens of the eye, which results in light scattering within the eye itself. Strong sunlight and reflectance from surfaces can cause temporary vision loss while driving. A recent driving simulator revealed that stray light had a stronger effect on driving performance than VA or contrast sensitivity among older drivers (Ortiz-Peregrina et al., 2020). Driving through partially shaded areas where there is intermittent strong sunlight followed by shade can prove problematic for older drivers as their ability to adapt to stray light is slower than younger drivers.

Glare recovery has been particularly difficult for older drivers. Glare, defined as brief periods of bright light (> 3000 cd/m²), strongly impacts photoreceptors in the eye, which, in turn, need time to recover to normal function. This is particularly so when driving under mesopic conditions. Under these conditions, drivers may experience strong and persistent after-effects. For those beyond age 65, recovery time is greater than younger drivers and increases rapidly with age. Median recovery time for those over 65 was found to be about 20 sec, while for those over 90 the median was 90 sec, with some requiring 160 sec to recover (Haegstrom-Portnoy et al., 1999).

VISUAL MOTION PERCEPTION

The basic visual sensory functions discussed thus far have broad implications for predicting the behavior of older drivers. How aging impacts a driver's ability to convert visual stimuli into useful information involves processes that occur at higher levels in the brain. An explanation of how movement, both of other objects external to the driver's vehicle and the vehicle itself, is processed by a driver was provided in Chapter 3. The effects of aging on motion processing are essential to understanding the behavior of older drivers. The processing of movement is divided into the motion of objects external to the vehicle on the one hand and the movement of the vehicle itself on the other.

OBJECT MOTION

One of the effects of aging on visual perception is the reduction in the ability of older adults to process the movement of objects in the visual field. Thresholds for the perception of object motion, particularly motions below 5 deg/sec, are higher for older than for younger adults (Bilino and Pilz, 2019). Indeed, the loss of sensitivity in discriminating object movement has been shown across all spatial frequencies and irrespective of contrast sensitivity for older adults (Raghuram et al., 2005; Snowden and Kavanaugh, 2006). These findings have implications for a variety of driving tasks.

LOOMING

One of the results of this poor discrimination for object motion is in the area of looming, which was discussed in Chapter 3. Objects approaching the driver's vehicle will appear to loom, a rapid expansion of retinal size, prior to collision. A study comparing older drivers with middle-aged drivers (40–55 years) found a strong negative correlation between looming test performance and age (Kennedy et al., 2001). Moreover, different rates of looming have been found to be reduced by between 2.8 and 3.4 mph for every decade increase in age (Poulter and Wann, 2013).

TIME-TO-COLLISION

For the specific task of estimating time-to-collision (TTC), older drivers perform more poorly than younger drivers (DeLucia et al., 2003). This poorer performance in TTC estimation is likely due to the problems of visual perception found in looming. A vehicle at greater distance subtends a smaller retinal size than closer vehicles and changes in that retinal size will also be smaller at greater distances. This puts the older driver at a disadvantage compared to younger drivers. In both looming and TTC estimates, the critical time margin necessary for collision avoidance is reduced. This critical time has been shown to increase with age beginning at age 50 and increasing substantially after age 65 (Uno and Horamatsu, 2001).

TURN GAP ACCEPTANCE

Turn gap acceptance necessarily involves TTC estimation, but also estimation of the time it will take to complete the maneuver. This is a typical maneuver at intersections where the left turn is not protected in a separate lane with traffic signals. The older driver needs to have a good estimate of TTC, the time to complete the maneuver, and an added safety margin. Studies of left-turn gap behavior reveal much larger gap requirements for older drivers than younger ones. In a study by Staplin et al. (1995), older drivers required a gap of an average of 22% more than younger drivers at lower speeds (e.g., 30 vs. 60 mph). Part of this is the larger safety cushion required by older drivers and the larger TTC estimate common in older drivers (Skaar et al., 2003).

MERGE

Merging into crossing traffic presents a similar problem, though not identical, to a conventional left turn as it requires a perceptual analysis of oncoming vehicles approaching at a heading perpendicular to that of the driver's vehicle. This presents a more difficult image processing task than vehicles approaching directly ahead. The rate of change in the retinal image would necessarily be much less per unit of distance. Even larger gap sizes were found for older drivers in a study by Yan et al. (2007). In this study, gap acceptance by older drivers was about 3 sec greater than younger drivers at 25 mph but dropped to only 1 sec longer when oncoming vehicles approached at 55 mph. The authors suggest that the drivers were more sensitive to the distance and position of the oncoming vehicle than its speed when making the gap acceptance decision.

FREEWAY ENTRY

Among the more difficult tasks that older drivers must perform, entering freeway traffic is high on the list. Freeway entry requires the ability to match their own vehicle's speed with other vehicles while estimating a safe gap between traffic for entry. This maneuver is done by judging both the speed of vehicles and the distance between them. Furthermore, these judgments are done relative to the driver's own vehicle movement. In an experiment on the effects of age on freeway entry, older drivers were found to merge at lower speeds and found the task more demanding than younger drivers (de Waard et al., 2009). The lower speed of entry by older drivers requires other drivers to slow down or execute evasive maneuvers in order to avoid collision.

SELF-MOTION PERCEPTION

Unlike the perception of object motion, the perception of self-motion is due to different visual cues. The perception of self-motion is heavily dependent on the sensation of movement produced by optic flow across the retina. When the vehicle moves, the optic flow increases in intensity as speed increases. Any reduction in the visual field of the driver for whatever reason will affect the sensation of the optic flow input to the retina. Thus, significant losses in the visual field due to aging will likely impact the perception of self-movement by older drivers.

A number of studies of older adults have found reduced sensitivity to optic flow patterns and associated reductions in self-motion perception for both speed and guidance (Warren et al., 1989). Reduced sensitivity of older adults has been found for both lamellar and radial flow optic flow patterns (Atchley and Andersen, 1998). Furthermore, inducing illusory self-motion or vection such as that which is provided in simulators appears to require higher levels of visual motion than younger drivers (Haibach et al., 2009). Reviews of motion and aging, however, show that older adults are more adept at integrating visual and non-visual cues (e.g., auditory, vestibular, somatosensory) than younger adults (Bilino and Pitz, 2019).

DISTANCE PERCEPTION

In an earlier chapter (Chapter 3), the judgment of distance between an observer and an object was attributed to the use of a variety of mostly mon-ocular distance cues. These visual cues include size–distance invariance of familiar objects, texture-density of the foreground, overlap of an object with another object, linear perspective, and other cues. How accurate dis-tance perception of objects is will depend on how well an older adult is able to use these visual cues. Distance perception is necessarily dependent on the perception of pertinent details of an object and the object foreground. Texture-density cues are effective only if the observer can discern the change in density of texture elements that occurs with increasing distance. Similarly, the size–distance cue is only useful if the observer can discern characteristics of an object that allow classification. It is likely, therefore, that issues of visual acuity and contrast sensitivity and other visual sensory functions will play a role in distance judgment just as they do in other tasks.

There is a dearth of studies investigating the role of aging on the per-ception of distance that is in the range important to most driving tasks (2 m or greater). One of these studies compared the ability of older adults, screening for visual anomalies, to accurately estimate the distances (4 to 12 m) of a fixed object viewed across a textured field in an outdoor setting (Bian and Andersen, 2013). The textured field is known to provide a strong cue to distance. The familiar object (a red brick) provided the size–dis-tance cue. In this study, younger observers underestimated their distance from the object. Older observers, however, were quite accurate in their distance judgments. In a second experiment, the authors manipulated both eye height and texture-density. The variation in eye height is important as it is known that eye height varies from one vehicle to another. Both vari-ables affected younger and older observers equally.

A more recent study of aging effects and distance perception was con-ducted using magnitude estimates of distance ratios (Norman et al., 2017). In this study, older and younger adults viewed pairs of horizontal poles each varying in length and judged the ratio of the distance between pairs at distances up to 55 m. Older male observers were more accurate than either younger observers or older females.

Finally, in a study by Dukes and associates (Dukes et al., 2020) the effects of age were evaluated in perceiving distance along the observer's line of sight out to a range of 36 m. Line-of-sight distance perception tasks the ability of the observer to discriminate increasingly small changes in detail with increasing distance. No evidence was found of any difference between older and younger adults in this task.

While more basic research is desirable in assessing the distance per-ception of older adults, the evidence does not support the assertion that older adults have inherently poorer distance perception due to the effects of aging. The question then is whether or not older drivers' perception of distance is comparable to younger drivers in a driving task where that ability is important.

DISTANCE PERCEPTION AND DRIVING

VEHICLE FOLLOWING

In normal traffic flow, it is inevitable that older drivers will be tasked with maintaining a safe distance from the vehicle in front of them. This following distance should provide sufficient time to bring the following vehicle to a stop before colliding with the lead vehicle. A common recommendation is for the driver to allow 3 sec travel time between the following and lead vehicles (California Driving Manual, 2001). At a speed of 30 mph, this translates into a following distance of 132 ft (40.2 m). This following distance allows sufficient time for the driver to respond to the onset of brake lights on the lead vehicle by moving from the accelerator to the brake pedal. While older drivers do not appear to differ from younger drivers in simple reaction time laboratory studies, the speed with which older drivers physically move their foot from the accelerator to the brake pedal has been shown to be slower than younger drivers (Olson and Sivak, 1986).

A study of older driver vehicle-following behavior compared the performance of these drivers with younger drivers on the ability to follow a lead vehicle delay (Dastrup et al., 2009). Older drivers had more difficulty matching the speed of the lead vehicle than younger drivers and tended to overshoot the lead vehicle's speed more than younger drivers when changes in the following distance were required. However, older drivers responded more quickly to changes in lead vehicle speed than younger drivers. Finally, older drivers maintained a greater following distance from the lead vehicle than younger drivers (82 vs. 44 m, respectively). The average following time for older drivers was about 269 vs. 144 ft for younger drivers. At a speed of 55 mph of the lead vehicle, this converts to about 81 ft per sec or 3.3 sec for older drivers compared to a 1.8-sec distance for younger drivers. Notably, the older drivers were maintaining a following distance in line with those recommended for safe separation, while the younger drivers followed the lead vehicle too closely.

The larger following distance maintained by older drivers necessarily results in a smaller optical image of the lead vehicle and, more importantly, requires a greater change in following distance in order to exceed the minimum rate of the expansion threshold of 0.003 rad/sec. The lead vehicle would tend to be displaced at a distance of 307.2 ft in 1 sec compared to a displacement of only 154.7 ft for the shorter following distance. This may explain why older drivers had difficulty matching the speed of the lead vehicle. The tendency to overshoot the lead vehicle by older drivers is likely a result of the older drivers' poorer motion sensitivity, particularly at the slow relative speeds of the two vehicles and the greater distances at which the lead vehicle would be viewed by older drivers. The result is an overestimate of the speed change in the lead vehicle. The more rapid response to changes in the speed of the lead vehicle by older adults is somewhat surprising. It is likely attributable to greater attention being paid to lead vehicle behavior by older drivers due to a greater risk-aversion for rear-end collision. This risk-aversion may account for the greater following distance, which is more than the typical 2 sec following time found with most drivers.

IMAGE CONTRAST EFFECTS

A more recent study of the vehicle following in older adults examined the role of varying amounts of roadway fog (Ni et al., 2007). Drivers were screened for VA, contrast sensitivity, and cognitive factors. Older drivers again revealed greater following distance when compared to younger drivers in clear conditions. As with the study above, older drivers again had more difficulty matching speed with the lead vehicle compared to younger drivers. The differences increased with lead vehicle speed and fog density.

Notably, the following distance *decreased* for older drivers as fog density increased. Reducing the following distance by older drivers when fog was added to the scene would seem to be counter to the risk-aversion expected from older adults in the vehicle following. However, fog impacts the contrast level of images to a point where the image becomes increasingly difficult to see as fog density increases. Reducing the following distance is a means of improving image clarity as the effect of fog on the lead vehicle image contrast is reduced with reduced distance. Reducing the following distance improves the image of the lead vehicle to the older driver despite the increase in perceived risk that the older driver will experience with the reduced following distance. It is also known that drivers close on a lead vehicle in fog as an aid to directional guidance (see Chapter 2) in curves. This helps replace other guidance cues that may be obscured by fog such as lane or road edge markings.

Older adults do not appear to differ from younger drivers with regard to basic distance perception ability. The differences that exist in vehicle following appear to be due to factors other than distance perception, per se. The greater following distance that older adults adopt in vehicle following appears to be a result of a risk-aversion strategy with regard to collision rather than a distance perception issue. The following distances found for older drivers were nearly a third greater than that of younger adults. However, this strategy has the unanticipated effect of making the detection of lead vehicle speed changes and speed matching more difficult. The addition of fog to the following task makes the problem of reduced guidance cues normally associated with fog much worse for older drivers due to their problems of motion perception at slower speeds commonly found in urban environments.

NON-VISUAL PERCEPTION

The three principal areas of non-visual perception of concern for older drivers are audition, proprioception, and tactile sensation. The sensory-perceptual functioning in these areas impacts critical operational areas within the driving domain.

AUDITION

Hearing loss with aging is a well-established phenomenon. Reviews of hearing loss rates show loss occurring at the rate of 1 dB every year after age 60 (Walling and Dickson, 2012). Hearing loss of more than 25 dB affects 37% of adults aged 61 to 70 and 60% or more of adults over the age of 71. Older drivers with moderate to severe hearing loss have been shown to be particularly vulnerable to auditory distractions while driving. In a study by Hickson et al. 2010), older, hearing-impaired drivers were found to have a significantly reduced driving performance in the presence of auditory distractors. The increased attention required of the older driver to process speech input appears to reduce that available to support driving tasks. Evidence of auditory distracters in the useful field of view suggests that hearing impairment in older drivers may have broader implications than previously thought (Herbert et al., 2016). Another study of older driver hearing loss revealed that older drivers will tend to shed tasks in the presence of a secondary task suggesting a strategy of reducing the load on an overtaxed system of attention allocation (Thorslund et al., 2014).

VESTIBULAR PERCEPTION

The vestibular organ consists of the semicircular canals and the otolith. The semicircular canals respond to head rotation and the otolith responds to linear movement of the head (either left and right or fore and aft). As with hearing, loss of hair cells in the semicircular canals can lead to problems in balance sensation among older adults (Anson and Jeka, 2016). However, it is also true that loss in vestibular response affects the Vestibular Ocular Reflex (VOR). The VOR plays a vital role in maintaining ocular stability when fixating on a target during movement. Lack of this stability is likely one of the reasons older drivers have difficulty reading road signage and may be implicated in problems detecting details of objects in the road environment.

The other vestibular component, the otolith, responds to translational movement. This would include responses to lateral accelerations which occur when a vehicle negotiates a curve. The translational motion also occurs when the vehicle accelerates or decelerates rapidly. The latter sensation occurs during emergency braking when the driver is responding. Research on older driver responses to translational motion is lacking.

PROPRIOCEPTION AND TACTILE SENSATION

Proprioception, the perception of the position and action of body limbs, is essential in the operation of vehicle controls. This applies to all controls but is particularly relevant to foot controls such as brake and accelerator pedals which are normally not in the direct view of drivers. These foot controls are normally operated by the driver positioning the leg and foot so they are aligned to the particular control. The position or angle of the ankle is also important when applying pressure to the pedals. Tactile or pressure feedback from the foot is also important in determining the pressure being placed on the pedal.

A review of the effects of aging on proprioception reveals impaired sensation with increasing age (Riberio and Olveira, 2007; Vershueeren et al., 2002). This decline applies to both lower and upper limbs as well as ankle joint position. Limb position detection thresholds increase with age as well. Ankle position sense declines from 3.4 to 6.5 deg, so much larger ankle position changes are needed to sense changes.

SUDDEN UNINTENDED ACCELERATION

The loss of proprioceptive and tactile sensation with aging may be implicated in sudden unintended acceleration (SUA). SUA occurs when a driver inadvertently confuses brake and accelerator controls, resulting in the application of an accelerator when braking is desired. Accidents resulting from SUA occur with younger and older drivers but are more frequent in the latter (Young et al., 2011). A study of pedal errors in younger and older drivers revealed higher pedal error rates among older drivers though both groups had similar brake reaction times (Wu et al., 2014). (Notably, error rates were two to three times greater for accelerator error rates than for brake pedal rates for both groups.)

HAZARD PERCEPTION

In reviewing issues involved in hazard perception in an earlier chapter (Chapter 4), the process of responding to a road hazard was divided into two categories: imminent or unanticipated hazards and potential hazards.

The differences in hazard categories are important in addressing the issue of aging as the two categories of hazard perception impact different physical and mental processes which are affected differently by aging.

IMMINENT HAZARD PERCEPTION

For imminent hazard perception where responses are required quickly in order to avoid collision or control loss, the responses involved are reactions to primitive sensory-perceptual mechanisms like accelerated looming. The responses by the driver do not involve conscious decision-making as usually defined but instead involve automatized visuomotor skills such as those involved in emergency braking and maneuvering.

For the older driver, the process of a simple reaction to a stimulus itself does not appear to be significantly longer than for a younger driver. However, the physical process of moving from the accelerator pedal to the braking pedal involves neuromuscular processes that make older driver pedal movement from the accelerator pedal to brake slower as well as more variable than younger drivers (Catin et al., 2004; Shin and Lee, 2012). Despite the slow movement from the accelerator to the brake pedal, brake PRT studies of responses to unanticipated hazards have failed to find differences between younger and older drivers (Fitch et al., 2010; Lerner, 1993; Owens and Sivak, 1986).

The physical movement of the steering wheel movement is significantly slower for older drivers with younger drivers steering wheel rotation nearly twice that of older drivers (Uno and Hiramatsu, 2000; Yan et al., 2007). These physical movement factors make timely maneuvering responses to imminent hazards more difficult for older drivers. The reduced sensitivity to the looming of potential hazards discussed earlier coupled with these physical limitations may make timely responses to imminent hazards impossible for some older drivers. The detection of potential hazards is, therefore, of much more importance to older drivers. The additional time afforded by detecting evolving potential threats or hazards to the vehicle generally allows for corrective action even with the physical limitations of older drivers.

PERCEPTION OF POTENTIAL HAZARDS

The perception of potential hazards, that is, the detection of hazards that may eventually cause a collision or loss of control, involves a complex process of visual scanning and search, detection, and classification. The perception may result in preventative actions (e.g., deceleration, lane change) to reduce exposure to the hazard. The potential hazard as defined in Chapter 4 is a hazard that may evolve with the passage of time and at a closer distance into an imminent hazard. This typically involves a collision with a static or moving object either already in the path of the driver's vehicle or an object that the driver suspects may enter the path of the driver's vehicle in the near future. Loss of control may occur if the vehicle leaves the roadway or if it encounters road conditions that may make vehicle control difficult or impossible (e.g., black ice).

Three key behaviors are required for potential hazard perception to be successful. First, as potential hazards can occur in many areas of the road environment, visual search and scanning behavior must be efficient to maximize the effectiveness of the search, particularly in complex visual environments. Second, the detection of potential hazards requires the ability to correctly classify objects, events, and environments that may represent potential threats to driver safety. Generally, training and experience in driving in a variety of road environments and traffic should provide the knowledge necessary to correctly identify these hazards. The development of situation awareness, which requires the ability to store and retrieve an array of potential hazards, is also necessary. Finally, the driver must develop skills to safely avoid these potential hazards without creating additional new threats in the process.

VISUAL SEARCH AND ATTENTION

Studies of the visual scanning of older drivers reveal deficiencies in the ability to efficiently scan the road environment. In a study comparing older and younger drivers, older drivers tended to concentrate on the center of the roadway and avoided scanning the sides of intersections where potential hazards, such as vehicles or pedestrians, may be entering (Yamani et al., 2016). In a separate experiment, when elderly drivers were required to search the roadside of a highway, the drivers needed more steering corrections to maintain lane positions. The narrowed visual scan of older drivers also appeared in an earlier study of visual searches at rural intersections (Bao and Boyle, 2009). A smaller number of visual fixations per second for older drivers were also found in complex intersections. However, a study comparing younger and older drivers using video clips found only small differences in scan path even though older drivers had a much higher false alarm rate (Underwood et al., 2005).

The results of these studies are in accord with studies showing a strong negative correlation between driver age and tests of visual attention (Richardson and Martolli, 2003). Visual attention scores were associated with a variety of real-world driving behavior deficiencies such as visual scanning and interaction with traffic and pedestrians. An index of visual attention, the size of the usable field of view (see Chapter 4), reduces with age and has been shown to be strongly associated with driving problems (Ball et al., 1993; Anstey and Wood, 2011).

The pattern of results for older drivers' visual scans in driving environments where attention may need to be devoted to multiple areas is a reflection of limitations in visual attention with increasing age. The effects of aging may also affect the ability to share or switch attention between basic vehicle control and the detection of potential hazards in the road environment as demonstrated in the study of elderly drivers discussed above.

HAZARD IDENTIFICATION

Less well understood is whether differences exist between younger and older drivers in how well they can detect hazards when visual factors such as contrast sensitivity are not involved (see Contrast Sensitivity and Driving section above). Part of the answer rests with the extent and type of driving experience that these respective groups may have. Generally, there is little reason to believe that experienced drivers differ markedly in their ability to correctly classify a threat once that threat has been detected. Nor is there evidence that older drivers will lose skill in identifying hazards. However, it is also possible that the differences that are seen in some studies of visual search are due to the fact that older drivers are more risk-averse and are focusing their visual attention on road components that they view as particularly hazardous while ignoring others. Studies comparing age and experience in hazard identification have not found significant differences between younger and older drivers although older drivers in hazard identification are somewhat slower in response to these hazards (Borowsky et al., 2009; Sciafla et al., 2012).

SITUATION AWARENESS

In an earlier chapter (Chapter 5), the situation awareness of the driver was described as a form of multitasking in which the driver needs to attend to a variety of driving elements including the identification and localization of potential hazards, the control of vehicle speed and lane position, and updating and retrieval of wayfinding information. As the identification of hazards is one component of situation awareness, that which affects the ability to maintain situation awareness will likely affect the perception of potential hazards.

One notable effect of aging is the reduction in the size and functioning of working memory (Salthouse, 1994). The reduction in the size and processing of working memory limits the ability of a driver to maintain multiple potential hazards in memory at any one time. It will likely affect the updating of navigation position information as well. As a result, even with competent visual search strategies, older drivers may be limited in their ability to store the location and type of hazards in more complex intersections. Some evidence of this can be seen in the behaviors of older drivers in complex traffic environments where the sensitivity of speed judgments of oncoming vehicles is reduced (Andrea et al., 2001). In a study of hazard effects in older driver SA, the combination of an increase in road complexity and the presence of dynamic collision hazards reduced older driver situation awareness (Kaber et al., 2012). Dynamic hazards appear to have a much stronger effect on older drivers than younger drivers possibly due to the risk-aversion response by older drivers. Probes of older driver situation awareness in the study revealed a decline in performance in SA elements related to road environment perception and wayfinding tasks. The latter is represented by increased missed turns (Trick et al., 2009). Attention to potential hazards is a key element in the driver's SA, but it is not the only element. The narrowing of this visual attention found in older drivers reduces the likelihood that other elements critical to SA will not receive adequate attention and, as a result, may be ignored entirely. SA is dependent upon the driver having the ability to share the focus of attention among a variety of road elements as well as having a working memory that functions to allow storage and retrieval of information. The development of SA may require, but it is not the same as, the ability to perform more than one task at a time.

MULTITASKING AND ATTENTION

The ability to focus attention and to shift that attention between different tasks is vital for safe and effective driving behavior. As discussed in Chapter 3, attention by experienced drivers is not normally required for highly automated visuomotor tasks like steering and brake or accelerator control. Attention is not required for reactive responses such as the collision avoidance response to an imminent hazard described above. These automated driving tasks are contrasted with controlled tasks which include those tasks which require attention. These include the visual search and detection of potential hazards, wayfinding and navigation decisions, reading and comprehension of regulatory sign signs, and other driving tasks that require selective attention.

Driving is an inherently multitasking domain in that it usually requires the dual tasks of vehicle steering control and vehicle speed control. As these two tasks are automatized in experienced drivers, they have little or no impact on driver capacity for selective attention unless an anomaly occurs. Multitasking involves the search for potential road hazards as well as some degree of wayfinding or updating of position with respect to the destination. A variety of other tasks relevant to driving may occur during a trip. All of these tasks are controlled in that they require focused attention. They may also compete for other driver resources beyond that of selective attention such as vision or audition or simply the use of a hand or foot.

To the extent that the physical and mental abilities of an older driver are affected by aging, older drivers' multitasking performance will be affected. In a study evaluating older adults' dual-task performance involving a compensatory tracking task and a visual choice reaction task, older adults revealed a decreased ability to divide attention between the two tasks when compared to middle-aged adults (Ponds et al., 1988). Large-scale reviews of dual-task performance by older adults show costs of adding a second task that goes beyond that found in younger adults (Verhaegen and Cerella, 2003). The additive costs were in the speed rather than the accuracy of performance and appeared independent of the complexity of the task. An earlier study by the same authors found the cost of switching between two controlled tasks was greater for older than younger adults (Verhaeghen and Cerella, 2002). These control processes involved in task switching may have been due to the addition of processing stages rather than the slowing of processing due to aging. Other reviews of aging and dual-task performance show an effect of aging on both accuracy and response speed of tasks particularly those requiring significant controlled processing (Riby et al., 2004).

AGING AND ACCIDENTS

Throughout this chapter, evidence was presented revealing the effects of aging on physical and mental processes and, in turn, the impact of those processes on driver performance. There has been little discussion of the older driver's diminished performance and its relationship to vehicle accidents. A separate section has been created here to address this issue due to the difficulty of drawing causal connections between a particular physical or mental deficiency and accident rates as accident data do not permit the kind of experimental controls that are needed. A second reason is that accident data are often missing data concerning these physical and mental processes. Moreover, large segments of potentially relevant data from low-speed, nonfatal, or low-damage accidents are never reported (about 30% of U.S. accidents). This data set is particularly relevant to older drivers who tend to avoid high-speed motorways and drive in more rural or low-traffic urban areas. In attempting to draw conclusions about older drivers and accident rates, it is important to note some key issues.

LOW MILEAGE BIAS

The first of the many problems with the use of accident data with demographic categories like age or gender is to establish whether or not a relationship actually exists. In other words, the likelihood of an accident can be determined solely on the basis of chronological age. A notable confounding of mileage with age was discovered by investigators who examined age in relationship to annual miles driven (Hakamies-Blomqvist et al., 2002). When accidents per annual miles driven were matched for each age group, no age relationship to accident rate was found. In a later replication of this study with more than 45,000 drivers, elderly drivers (75 or over) actually had lower accident rates (Langford et al., 2006). The only case where accident rates were higher was for those driving less than 1800 miles (3000 km) or less per year. This consisted of about 10% of the study sample. It appears that as the driver's age, the various physical and mental declines, and their impact on performance as well as the risk-aversion tendencies of older drivers may result in dramatic reductions in miles driven. Despite the low mileage driven, however, these deficiencies can lead to higher accident rates than in other age groups. The question then is why this particular group is experiencing higher accident rates.

MEDICAL IMPAIRMENTS

A variety of impairments may account for the high rate of accidents in the older-driver, low-mileage group. A study of 4,316 older, low-mileage drivers were evaluated for fitness to drive as defined by their medical, visual disease, and mental health test results (Alvarez and Fierro, 2008). Those who failed these tests were deemed "unfit to drive", but some were classified as "fit to drive with restrictions". The accident rate of the latter was monitored for the following year. There was no significant difference in accident rates between these and other low-mileage, older drivers. Medical impairments, as such, do not appear to be the reason that low-mileage, older adults have unusually high accident rates.

PERCEPTUAL AND COGNITIVE IMPAIRMENTS

The absence of medical impairment as predictors of the low-mileage, older-driver accident rate suggests that the reason for accident rates in this group is more likely due to other factors that may relate to the sensory, perceptual, or cognitive abilities of the older drivers. A more recent study by Langford and associates (2013) examined the sensory, perceptual, and cognitive components of the fitness to drive testing and found that these abilities, unlike medical tests, did in fact predict accident rates in this group in the ensuing 12 months of driving. These included measures of balance, an indicator of vestibular function as well as VA, and measures of cognitive ability (e.g., attention).

We know that older drivers have more difficulty dealing with potential hazards. Their slower responses to these potential hazards have been shown to be related to higher accident rates (Horswill et al., 2010). Elderly drivers have been shown to be over-represented in rapid deceleration events (RDEs), which are also associated with accidents. These RDEs were correlated with losses in contrast sensitivity in the previous 12 months (Chevalier et al., 2017). Furthermore, the losses in working memory size and function contribute to difficulties in situation awareness, which is vital to safe driving in complex intersections. Thus, the low-mileage, older-driver's ability to drive is likely due to a constellation of sensory, perceptual, and cognitive process declines, which impact critical skills including the ability to accurately estimate time-to-collision and to detect, identify, and respond to potential hazards. The result is an increase in the accident rate for this group.

DRIVING ENVIRONMENT

Low-mileage, older age groups as defined above are more likely to drive in urban environments where exposure to unprotected left turns and complex intersections are more likely to occur. Older drivers tend to be risk-averse to the higher-speed traffic found on freeways and expressways traveled by younger drivers. As noted earlier in this chapter, left turns and complex intersections present a significant challenge to older drivers. Moreover, urban settings will expose older drivers to more traffic conflicts than rural roadways; the greater the number of traffic conflicts, the greater the accident rate. A review of accident studies indicates that low-mileage, older drivers are much more likely to be exposed to a complex, urban environment, and this, in turn, results in more accidents (Mayhew et al., 2006). In contrast, the low mileage bias in older drivers does not exist when only rural accidents are studied (Hanson and Hildebrand, 2011).

SUMMARY

This chapter reviewed the evidence for losses in sensory, perceptual, and cognitive function associated with normal, healthy aging. Each of these functions contributes in some way to the process of competent driving behavior. While VA, the most commonly tested of the sensory functions, declines with age, the average VA functionality is sufficient to perform most driving tasks well into old age. However, certain driving-related tasks such as sign reading, hazard detection, and object motion processing are affected by the loss of contrast sensitivity. This loss of contrast sensitivity begins at age 50 and continues to decline for the remainder of life. Loss of contrast sensitivity is particularly problematic for driving in low luminance conditions in the daytime and especially under mesopic luminance conditions at night. Losses in auditory, somatosensory, and vestibular function also occur with age. Their impact on driving behavior is less serious than other sensory dysfunctions but may be a contributing factor in reduced driver performance when multiple tasks are required of the driver. Perceptual functions such as distance perception and self-motion perception do not appear to be affected by aging.

The low-mileage, older driver cohort reflects the diminished capacity of some older drivers to perform the driving task who then reduce their mileage as a result. This is apparently due to a variety of sensory, perceptual, and cognitive factors that decline with increasing age. For some older individuals, this decline means that they are no longer comfortable with long trips and prefer shorter trips in less demanding, urban environments. However, this exposes them to complex intersections and left-turn maneuver requirements for which these same diminished capabilities often result in collisions. Due to the lower speed when compared to higher-speed roadways, older drivers would nonetheless have a higher likelihood of survival in such urban accidents.

7 Resource Management and Driving Behavior

For complex, dynamic environments like those in which driving occurs, the physical and mental resources of the human operator, including their skills and knowledge of a given task, are often taxed to their limits and beyond. Even in routine driving activities, drivers need to manage resource expenditures in order to have at least some remaining resources to respond to emergent hazardous events. In this chapter, the resource management framework will be elaborated with particular attention to the driver's ability to prioritize, interleave, and shed as well as use other compensatory behaviors to manage resources. Vehicle design, including automation and information display design, and the design of the road environment are discussed in terms of how each can serve driver resource management. Before this discussion, a review of some of the prior attempts to understand driver behavior will help place the resource management framework with respect to other attempts at a broader understanding of driver behavior.

In the previous chapters, driver resources were examined in relation to primary driving tasks including basic vehicle control, collision avoidance, and hazard perception. This view is in contrast to previous views of driving behavior which have focused on driver risk behavior (Wilde, 1998) and the relationship between driver task demand and driver capability (Fuller, 2005). The principal objective of these theories is the understanding and prediction of driver behavior particularly with respect to accident rates, especially collision risk. Theories based on driver resources such as Wickens (2002) focus on how driver perceptual resources such as vision and audition function in dual-task environments. These three main rhetorical approaches are only a selection of those theories and models available. The intent is to show the range of theoretical and modeling attempts at explaining driver behavior.

RISK HOMEOSTASIS

The fundamental tenet of risk theory is that improved accident rates in driving can be achieved only if drivers are willing to reduce the amount of risk they are willing to take (Naatanen and Summala, 1974; Wilde, 1998). In Wilde's risk homeostasis theory, the amount depends on four factors: (1) the benefits of risky driving behavior, (2) the costs of that behavior, (3) the benefits of safe driving, and (4) the costs of safe driving. Drivers adjust their driving behavior in a given situation according to a pre-determined, target risk level. If, for example, driver speed is perceived as exceeding

DOI: 10.1201/9781003454373-7

the perceived risk under the current prevailing conditions, the driver will reduce speed to match the level that meets the targeted risk level. Risk homeostasis is dynamically controlled by the driver under each driving condition, which results in a stable, predictable outcome.

For driving in this risk-management manner, the benefits of reduced driving risk must be balanced by the cost. In this case, the lowered risk reduces the probability of collision or loss of control and threat to bodily injury or property loss. The cost of this behavior will likely increase the travel time to the destination. The travel time increase to the driver must be justified by the reduced risk to maintain the benefit-to-cost balance. The balance or homeostasis of these factors is essential to reducing accidents in the risk theory of driving.

TASK DEMAND AND DRIVER CAPABILITY

A response to the risk homeostasis theory of driving behavior was developed by Fuller (2005; 2011). In this view of driving behavior, the risk component of risk homeostasis is subsumed in a model centered on the interaction of driver task demand and driver capability.

In the task demand capability interface (TCI) theory, the driver's subjective state homeostasis is again postulated as the controlling mechanism for behavior just as it is in risk homeostasis theory. The essence of the TCI theory is the dynamic nature of task demand that the driver perceives at any one time and the driver's perceived capability or skill to deal with that demand. It is the interaction of driver task demand and capability that defines task difficulty in driving. Accordingly, an individual driver will "adapt a level of task difficulty they wish to experience when driving" (Fuller, 2005, p. 462). To reduce task difficulty, drivers will reduce task demand until it matches capability. As both task demand and capability are subjectively perceived by individual drivers, there are no objective determinants of task difficulty. If task difficulty is felt to be too high, then task demand will be reduced. If it is too low, then an increase in task demand may be sought to more readily match capability. Driver capability in this theory is constrained by the biological characteristics of motor coordination, information processing speed and capacity, and reaction time. Driver skill and knowledge acquired through training and experience combine with these factors to set the upper boundaries of driver capability. Driver capabilities may be compromised by fatigue, ill health, motivation, and other factors.

Driver Task Demand

The demands of the driving task are a function of a variety of factors including road environment design elements such as road geometry, road surface conditions, road markings, visibility, and similar factors. Task demand in driving is also affected by traffic density and the behavior of other drivers.

The driver's vehicle itself affects task demand. Vehicle control characteristics in speed and steering impose task demands that may exceed driver capabilities under certain conditions. In-vehicle technologies can affect task demand as well. Task demand decreases with the automation of control tasks, for example, and will increase task demand when multiple, concurrent tasks demand attention.

In a given driving task such as steering control through a curve, the TCI theory predicts that a driver of a given level of capability will perceive the task demands of this task in relation to the driver's assessment of their own control skill. The resultant comparison will determine whether the perceived task demand is in excess or below capability. If in excess, the driver will reduce speed to reduce task demand and thus task difficulty. Thus, achieving TCI homeostasis by reducing task demand also acts to reduce risk and the possibility of exceeding driver skill and possible loss of control. The continual comparison of task demand and capability by the driver leads, according to the TCI theory, inevitably to the reduction of accident rates. In TCI theory, the driver's role is to adapt to the task demands that a given driving situation imposes. The principal method of adaptation is the reduction of speed through other means, such as task shedding, are available.

Managing task demand to match driver capability in this self-regulating way would account for compensatory behaviors such as shedding of a task in high workloads or changing routes to reduce exposure to more demanding traffic environments. Shedding of a given task in the TCI model is only done in relation to the capability of the driver to perform driving tasks. Nothing in the theory provides for prioritization of one task over another. Drivers shed tasks based on their individual effects upon overall task demand. However, in the absence of any means of prioritizing the effects of a given task in advance, the impact on demand in multitasking can only be determined *after* it has been shed. Without some means of prioritizing tasks, the driver is left in an endless trial-and-error process of adding and shedding tasks until a balance between task demand and capability is re-established.

MULTIPLE RESOURCE THEORY

Both risk and TCI homeostasis models depend on a driver's subjective assessments of risk, task demand, and their own individual capabilities. The simplicity of these "top-down" approaches to driving behavior makes them attractive to a driver-centric view of accident causation. However, the limited resources and capacity of the driver to manage a complex task domain such as driving are not considered in such views. The driver is required to adapt to the prevailing driving environment regardless of vehicle design, traffic conditions, or road environment issues. The lack of consideration of these driving domain issues in a theory of driving behavior cannot account for the influence that design issues inevitably have on driver behavior and on accident rates as the driver is inevitably at fault for any driving outcome regardless. Placing the full burden for safe driving on the driver is not justified or justifiable as the safe performance of a vehicle is a collective function of vehicle design, road environment, and driver resources and capabilities. A broader approach to driving behavior is needed to assure that the driver–vehicle interaction with the road environment is fully understood.

Homeostatic, top-down views can be contrasted with "bottom-up" views that examine driver resources at the perceptual level. This view was developed to address the competition among resources that comes with behavior in complex task environments such as driving. How the driver manages resources in multitask conditions is an important element in understanding driver behavior in general.

Consideration of how the vehicle and road environment affect the driver is necessary for a complete understanding of driver behavior. For this reason, alternatives to the driver-centric view of driving behavior as described above have been put forward. One such approach is to examine the driver's perceptual and cognitive resource utilization in meeting driving task demands. The multiple resources theory of Wickens (2002, 2008) and Horrey and Wickens (2003) was developed specifically to address the problem of resource use in predicting dual-task interference. This theory and the computational model derived from it address the issue of interference between two tasks which are carried out concurrently. Assessment of differences in performance, when different tasks are performed at the same time, was evaluated specifically for tasks which shared visual or auditory resources though the model could be extended to other sensory modalities as well.

In the theory, dual-task performance is defined in reference to four dimensions. The competition among these dimensions determines how much interference or cost will be incurred by, for example, the driver of a vehicle. The four dimensions are *Stages*, *Processing Codes*, *Input Modality*, and the *Visual Channel* (focal or ambient). Stages refer to information processing stages: perceptual and cognitive processing as one stage and response processing as the other. Processing codes refer to the analogue codes associated with manual control and symbolic codes typically associated with language processing. The input modalities refer to the visual and auditory systems of the driver. The visual channel is divided into focal vision and ambient vision, which have been discussed in earlier chapters. Resource competition at each of these dimensions is a result of task sharing, which is likely to have a negative effect on performance.

A validation study of this model of dual-task interference was conducted by Horrey and Wickens (2003). A driving simulator study of resource competition was evaluated in a variety of rural and urban scenarios for the purpose of determining the predictive validity of the model. An in-vehicle visual and auditory display system provided the secondary task stimuli. Drivers were instructed to maintain safe vehicle control and avoidance of road hazards.

The resource competition model predicted most of the variance in performing the secondary in-vehicle display task and nearly all of the variance in driver response time to road hazards. The model did not predict variance in lane keeping although it was a primary task.

In this study, interference was revealed principally in degraded in-vehicle task performance and in hazard detection. Performance decrements occurred in hazard detection despite an instruction to the driver to prioritize hazard detection.

While the detailed computations of the resource competition model are beyond the scope of this book, a number of important elements are of interest in this approach to the problem of dual-task interference. First, the demand for limited resources is particularly acute when those resources are needed for consciously controlled tasks. These include the use of focal vision and the processes involved in tasks such as hazard perception. Lane position tasks are controlled primarily by the use of pre-attentive ambient vision. As such, it would not compete for higher-level resources needed to carry out either the in-vehicle display task or the hazard detection task as both rely on focal vision. (Note that focal vision is not a sharable resource for other tasks.)

Multiple resource theory, unlike the two previous theories described, may be categorized as a "bottom-up" theory of resource utilization. While theories centered on risk or task demand and driver capability attempt to account for the substantial contribution of human behavior to accident rates, multiple resource and similar modeling activity focuses on the more practical applications of human performance modeling. This is especially applicable to the issue of in-vehicle display system design. However, resource models need to be expanded to address input modalities beyond vision and audition and the visual channel to include other inputs including the sense of touch as well as focal audition. These resource areas will see increasing use in the design of in-vehicle technologies, particularly haptic technologies that apply force and vibration to the driver, and the increasing reliance on auditory displays to reduce dependence on driver focal vision. Additionally, the problem of resource allocation among tasks and task prioritization remains difficult to predict and explain within the context of a broader scope of resource management in driving behavior.

RESOURCE MANAGEMENT

In the complex multitasking environment within which drivers operate, their sensory, perceptual, perceptual-motor, and cognitive resources are often taxed to their limits and beyond. It is also true that driving task demands reach such low levels that the driver's attention drifts away from driving to other non-driving distracting tasks. Thus, the level of task demand can vary widely on a given driver's route, requiring the driver to be at a level of situation awareness that must be sustained for long time periods. Moreover, drivers often need to manage resources distributed in time over more than one driving task as well as manage the impact of any distracting tasks such as cell phones and passenger queries.

Approaching the driving task domain by characterizing it as a resource management problem does not necessarily mean that all resources need to be managed consciously by the driver or that all resources are even under conscious awareness. For example, both visual and non-visual sensory and perceptual systems operate in driving tasks without conscious awareness as they would in other non-driving tasks. The various low-level sensorimotor activities involved in walking, in sports, and even in skilled manual labor occur with little or no conscious awareness. This is partly due to biological development and partly due to concentrated skill development that automatizes these motor skills through continued repetition. Skill sets developed in this way operate within a particular set of operating parameters. Thus, a driver's ability to visually perceive visual motion, perceive distances, hear sounds, and sense physical motion does not change much through a driver's career until well into old age when these mental and physical resources begin to functionally decline. Barring ill health or drug use which debilitates the central nervous system and motor response, the processing and response speed of drivers to road events are well within the range of predictability by modern science.

Our ability to manage these driver resources effectively is a function of a variety of methodologies including, but not limited to, driver skill development and automatization, driver resource management strategies, and driver compensatory behaviors to manage demand on driver resources. Apart from driver strategies and skill development, resource management is also an issue in the design of in-vehicle technologies as well as the automation of vehicle control. Finally, the design of the road environment including signage, road geometry, intersection design, lighting and traffic flow, and other elements can aid in driver resource management of the driving task.

AUTOMATIZATION AND SKILL DEVELOPMENT

Automatization of driver perceptual-motor skills has been discussed in earlier chapters. However, the process of automatization and its current and potential impact on driver resource management is deserving of further discussion. Automatization of these motor skills represents a means by which driver resources, especially attention, can be conserved for use in other areas of driving. Visual attention resources can be diverted to the detection of potential vehicle and pedestrian collision hazards. Automatization in theory can occur for any driver motor skill and is, therefore, a key element in driver resource management. The importance of automatization of skills in driver training is essential and should be emphasized in driver training and testing programs.

The process of automatization of skills appears to be a function of a number of factors. First, the relationship between a response and a percept must be direct and specific. A simple control response of steering input or steering angle is only in relationship to a particular visual input. That visual input is typically the tangent point in road curvature for steering control. In the case of emergency braking, it is the rapid expansion of an object as an imminent collision threat. This does not preclude the same action being used in response to other stimuli but the response will generally require consciously controlled, perception-action sequences. Thus, steering in the process of maneuvering a vehicle into a parking space or braking as a part of the normal deceleration into an intersection are both sequences that require consciously controlled actions. These perception-action sequences are necessarily more demanding of driver resources, particularly focal vision and the perception of closure rate.

The origins of automatized driving skills are not necessarily solely in repetitive driving responses. Steering control actions are common in other vehicles that novice drivers may be exposed to such as bicycles or motorcycles. The steering control response to change vehicle direction is then likely imported into motor vehicles and refined through training and experience. A similar process likely occurs in the reactive response to an imminent collision. Reactions to objects on a collision course occur with pedestrians negotiating the driving environment as well as in other activities in life. This instinctive reaction to a potentially harmful or even lethal threat is so pervasive as to make it likely to be pre-wired in humans and other animals as evidenced by the existence of looming detectors in the brain. In driving, automatization occurs in the estimation of time to collision with the object to the vehicle control avoidance response, either braking or maneuvering or both, to the impending collision. Automatization occurs during driver training as well as during a driver's operational career. It typically occurs without explicit training but rather as incidental to the process of overall driving skill acquisition. The precise time or number of repetitions required to automatize vehicle control skills in driving is not known. Driver training is generally done in the whole task; e.g., control skills are trained at the same time other non-control skills such as hazard perception and collision avoidance are trained as well.

Some control skills, due to their complexity, may take longer than others to achieve an automatized state. Manual or standard transmission shifting requires relatively complex series of actions when compared to steering or braking. Initiated by a perceived need to increase or decrease vehicle speed, the shift sequence begins with removing the foot from the accelerator and then disengaging the clutch by depressing the clutch pedal. The driver then shifts to a new gear followed by the release of the clutch pedal. During the learning process, the driver may have to look at the shifter to execute the shift though this need disappears with automatization. This may be followed by pressing the accelerator pedal if acceleration is desired. The manual shift task components are designed in a hierarchy where a *structural prioritization* of task components is enforced to assure that the task goal (gear change) is achieved. Tasks like manual shifting are termed *procedural skills* as they require actions to be performed in a specific sequence. Repetition during training and field experience of the task of manual gear shifting results in overlearning and automatization.

Structural prioritization means that driver resources will be allocated to each task component in accordance with the needs of the particular task in order that the task goal will be achieved. The greater the contribution of the task component is to the achievement of the task goal, the greater the value the task component has and the more likely it will be executed. Once the decision is made by the driver to change gears, the sequence of subtasks is executed automatically beginning with the clutch pedal depression.

Complex sequences in perceptual-motor skills like manual gear shifting take longer to overlearn and, therefore, to automatize. Automatization means that very little or no attention is needed for manual shifting skill execution; as an automatized skill manual shifting is much less likely to interfere with other driving skills that may be needed at the same time as a critical resource, driver attention is no longer required. In road signage detection, drivers with many years of manual shifting experience were just as proficient in road sign detection tasks as drivers with comparable years of driving with automatic transmissions (Shinar et al., 1998). However, novice drivers averaging about 1 year of manual shifting experience were poorer in this task than novices with comparable experience with automatic shifting. In this instance, the attention demand required of manual shifting interfered with novice driving performance even after more than a year of experience. Although automatization of driver motor skills such as shifting, steering, and braking are all likely to occur eventually during a driver's career, more information is needed on whether there is a specific need for more training on these types of skills before licensure.

An additional issue of automatization of driving skills is whether it lasts for an entire driving career. Reviews of motor-skill learning indicate, although the capability to learn motor skills remains intact with age, motor-skill performance tends to decline with age (Voelcker-Rehage, 2008). With respect to driving skills such as manual shifting, a decline in driving performance might be expected if the decline in automatization requires more attention to be paid to shifting than for younger drivers. A driving simulator study comparing older with younger drivers showed that older drivers were involved in more collisions than younger drivers, but only when manual shifting was required (Piersma and de Waard, 2014).

AUTOMATIZATION OF COGNITIVE SKILLS

Thus far only those driving skills that involve motor skills, such as those requiring manipulation of hand or foot controls that involve neuromuscular components, are the only driving skills known to be subject to overlearning and automatization. Even driving skills that qualify for automatization need to have a strong and specific relationship between stimuli and response. A simple light onset, followed by a press of a switch controlling the light, represents such a relationship. Steering wheel thumb controls used to control the volume of a radio are another example of such a relationship. Such simple relationships between hand or foot controls and a definable stimulus have the potential to reduce or eliminate the impact on driver attention particularly for in-vehicle control and display devices.

While it is evident that driver motor skills are subject to automatization, the question is whether other skills not involving motor components would be amenable to automatization as well. Evidence for successful automatization of a skill is that the skill no longer requires the level of attention it needed prior to overlearning. An overlearned skill will be much less vulnerable to interference from a secondary task which requires attentiveness such as cell phone use. Studies of interference of skills involving basic perceptual processes such as gap acceptance tasks in left turns were disrupted by a secondary task involving checking the accuracy of sentences over a telephone connection (Brown et al., 1969). In another study, following distance from a lead vehicle was found to be unaffected by a secondary task (random letter generation) whereas braking for an intersection approach and mirror checking were affected by the task (Duncan et al., 1992). This provides evidence for the possible resistance to interference by some primarily perceptual tasks such as vehicle following and the vulnerability of other tasks such as dynamic estimation of changing distances and components of some complex maneuvers such as overtaking.

Driving hazard perception, the ability to detect potential road hazards or threats including collision or loss of control, is a cognitive, not a motor, skill. It primarily involves visual search, detection, and classification of objects and conditions on the road ahead. As a cognitive task using focal vision and cognitive processes, it is heavily dependent on driver attention. If the task could be automatized through overlearning, the impact on driver cognitive resources could be lessened. A study by MckEnna and Farrand (1999) compared the performance of novice and experienced drivers on a hazard perception test in the presence or absence of a secondary task (random letter generation). While experienced drivers were more proficient on the hazard detection test than novices without the secondary task, both groups' performance in hazard detection declined when required to perform the secondary task. This indicates that hazard perception is vulnerable to interference even to experienced drivers with more than 10 years of driving experience.

TASK MANAGEMENT

Automatization of certain tasks is perhaps the optimum means by which the driver resource of attention can be managed. However, drivers are known to develop strategies to deal with task demand that exceeds their available resources. One of these strategies is to abandon a subtask, task element, or an entire task in order to unload demands on overtaxed resources. A mechanism of driver task prioritization is needed in order for the driver to reduce task demand by means of task abandonment or shedding. It was noted that within a specific task, an element of a task can have value based on its importance in meeting a task goal. That is, it is structurally prioritized when the skill is developed. For example, eliminating the use of a rearview mirror in passing is more acceptable than visual attention to the oncoming vehicle due to the higher structural importance of the latter to the former in reaching the task goal. This prioritization comes with training and is reinforced with experience in the execution of a particular task.

In the case of multiple tasks, the order in which the tasks are executed appears to affect the decision as to which will be abandoned first even when one task should have priority over another. In a study of task prioritization in a following collision avoidance task by Levy and Pashler (2008), drivers were presented with concurrent tasks with either the following task or a choice response to an auditory task. Drivers were instructed to always consider collision avoidance as the priority task and to ignore the other tasks entirely. The results showed that drivers, despite instructions, tended to respond to the low-priority task with the collision avoidance task suffering as a consequence. The failure to withhold responses to the low-priority task occurred because the cognitive processes involved in the decision-making for this task interfered with those processes involved in collision avoidance. Only the addition of urgency to collision avoidance, resulting from the addition of brake lights to the lead vehicle, offsets the influence of the order in which the responses to these tasks occurred. The interference of a low-priority task on a high-priority task may occur if the processing of the response to a low-priority event occurs with close temporal proximity to the response to the high-priority event. Shedding of a low-priority task may not be possible if it occurs in close temporal proximity to a high-priority task even if that task involves vehicle safety.

Perhaps the most common means of driver task management involves prioritizing tasks in accord with a value assigned to the task itself by the driver. In the driving task domain, driving tasks have inherent value in supporting the ultimate driver task of safe transportation from one point to the other. In order of objective task importance, collision avoidance, vehicle direction and speed control, detection of potential road hazards, and navigation and wayfinding should correspond to the value or importance to the driver in defining task priorities. These essential driving task priorities should be inculcated in the driver during training and reinforced with experience. In turn, any of these essential driving tasks will be given priority over any distracting task regardless of the latter's perceived importance to the driver. Unfortunately, this ordered prioritization of task values both within the driving domain and in its primacy over non-driving tasks is lacking in many drivers. The result is that low-priority tasks may demand resources required for essential driving tasks, such as collision avoidance, with accidents as a potential result.

Incorrect task prioritization can actually induce task mismanagement. For various reasons, some non-driving tasks have inherent value for the driver even though they may demand resources needed for a driving task. In the case of cell phone use, for example, the response to a cell phone ring may be limited to ignoring the call if the driver believes answering the phone is of low priority compared to the driving task at hand. However, the driver may be expecting an important business call or have some other reason, which elevates the value of the call to a higher status. The precise timing of cell phone call reception is unpredictable so the strategy of using the cell phone may inevitably result in conflict with a driving task. Young novice drivers often rely on cell phone use for social communication in both texting and non-texting forms. The reliance on the cell phone at this level of social interaction increases substantially the value associated with its use. Ignoring incoming calls from friends is difficult at best even when driving task demands on driver resources are at a maximum.

Similar issues of task prioritization occur with other sources with the vehicle. Passenger communications remain a source of in-vehicle demand on driver resources. Passenger conversations with the driver require cognitive resources required for speech reception, speech production, and working memory to sustain a conversation at any length. Unlike cell phone use, simply ignoring or shedding a passenger conversation may not always be possible given the strong social importance of communication, particularly with family and friends.

Passengers, especially those who are licensed drivers, can be an important driver resource. That resource can extend to augmenting driver visual search for potential road hazards as well as aiding in navigation and wayfinding tasks. Unlike cell phone communications which are remote, the driver has much more control over the pacing of passenger communications so that the communications can be integrated with driving tasks rather than competing with them (Drews et al., 2008). Passengers can also exert supervisory control over some drivers, such as teenage drivers, or drivers who are family members. In this case, passengers can enforce safety standards such as speed control and other issues which may result in general driving skill improvements.

Related to the pacing of communications between driver and passenger, another means of managing driver resources is for the driver to interleave tasks so that visual focal attention is alternatively switched from one task to another. This is a particularly common strategy used by drivers to switch between externally focused tasks such as steering or speed control and in-vehicle tasks. Studies of how drivers interleave different tasks are informative with regard to how resources are managed dynamically. Complex secondary tasks such as cell phone use are often candidates for interleaving with driver tasks because they occur over longer time periods. This means that the driver cannot complete the secondary task without an extended period of time away from the driving task. In the case of cell phone use, the dialing of the phone may require the manual entry of as many as a large number of digits. The intuitive assumption would be that the driver would wait until a natural break in the digit sequence to switch back to the road. A study which examined driver behavior in vehicle lateral position control and dialing found that drivers do not interleave at the natural break points in dialing. When prioritizing driving tasks, drivers rely on vehicle lateral deviation limits as indicators to switch visual attention from dialing to the road (Janssen and Brumby, 2010; Jansen et al., 2012). The drivers are apparently attempting to balance attention to each task while maintaining the priority of the steering task. Importantly, time away from the driving task may be the more important determinant as drivers will tend to interleave if the length of the natural break in a secondary task is judged to take too much time away from the road.

HAZARD PERCEPTION TRAINING

While the hazard detection task may involve repeated exposure to drivers of the same hazard, the requirement of a direct-stimulus response relationship for automatization would appear to be met. However, it is possible that tasks such as hazard perception, which do not have a motor component as a primary element of the task, are not subject to automatization for that reason. It is possible that hazard perception need not wait for the driver to develop the skill through on-road experience, however. There is evidence that hazard perception part-task training can improve the performance of a driver's ability to detect hazards. This has been demonstrated for both experienced (Crick and McKenna, 1992; Horswill et al., 2013) and older drivers (Horswill et al., 2015).

A variety of hazard perception training programs have been developed for the training of teen and novice drivers. These include the Engaged Driver Training System (EDTS), Accelerated Curriculum to Create Effective Learning (ACCEL), and Risk Awareness Perception Training (RAPT), among others. Within the context of resource management, these training programs facilitate the development of awareness and attention to hazards in the road system. Detection and discrimination of these potential hazards improve the overall efficiency of resources such as visual scanning and the focal visual attention necessary to detect and classify potential hazards. Even relatively simple, tablet-based hazard perception training programs for teens have been shown to be effective (Ahmadi et al., 2018).

The key to improving driver resource management is, in the short term, improving driving skills through concentrated training in skill automatization and in areas such as hazard perception. Additional instruction should be provided in the prioritization of tasks within the driving domain and eliminating distracting tasks which expend important driver resources like focal visual attention on non-driving activities.

AUTOMATION AND IN-VEHICLE TECHNOLOGIES

A discussion of driver resource management would be incomplete without a discussion of the role technology is now playing, and will play, in how drivers interact with both the vehicle and the road system. The role of technology in other areas of transportation such as civil aviation has been to enhance operator resources, on the one hand, and to simply replace the operator on the other. In civil aviation, pilot aids such as traffic collision alert systems aid the ability of the pilot to detect and avoid collision hazards. On the other hand, automation has resulted in the pilot being effectively eliminated from continuous, active control of the vehicle and instead moved to the role of supervisory control. In this role, the pilot is a monitor, not an operator, of aircraft systems. Automation is now being applied to many of the tasks previously done manually by drivers including vehicle speed and directional control, and collision detection and response. The continuous monitoring of the vehicle state and its surrounding environment means an increasing reliance on alerting and warning systems as the driver will likely be unable to monitor the state of all systems continuously for any length of time.

The application of automation and supporting technologies to the driving environment should not assume that the consequences for driver resource management and vehicle safety will follow the same path as aviation. Commercial aviation has achieved a high degree of safety and efficiency with automation, but the safety achieved is also due to the rigors of pilot training, testing, and recurrent training. The airline pilot is a highly trained professional under the supervision of other professionals operating aircraft systems which are maintained at a high level of readiness. The typical driver, barring unusual events, is not usually subject to retraining or even testing nor is the vehicle operated under such scrutiny. Automation technology needs to be self-monitoring to assure operating parameters and sensors are within specifications. Drivers will need to be alerted when these systems are not operating properly and, where necessary, simply disengaged.

Automation of the vehicles that drivers are likely to operate in the near future fall into three general categories: full manual control where all aspects of vehicle control including collision avoidance are controlled by the driver; semi-automated or partially automated in which some vehicle functions are under driver control; and fully automated where all of vehicle control tasks previously performed by the driver are now automated. A variety of schemes to further delineate levels of automation have been put forward. The most common of these in the U.S. has been that put forward by the Society of Automotive Engineers (SAE, 2021) in which automation is divided into five levels ranging from 0 (no automation) to Level 5 (full automation). Driver assistance systems (Level 1) such as adaptive cruise control (ACC) are considered the first step toward automation but are not themselves considered automated systems as they are only engaged momentarily. While such systems may not be technically considered automation, the impact they have on driver resource management is of concern as these systems may create driver dependencies over time such that driver resources are no longer expended for the tasks these systems perform. Those resources are likely to be spent on other tasks, whether driving-related or not.

DRIVER ASSISTANCE SYSTEMS

A variety of systems aimed at assisting the driver termed advanced driver assistance systems (ADAS) have been designed with the intent of aiding the driver in the performance of certain tasks. Among the systems included in ADAS are ACC and lane departure warning (LDW) systems. ACC is a system which, when engaged by the driver, will maintain a specific speed and will adapt that speed when approaching a lead vehicle in order to provide a safe distance between the driver's vehicle and the lead vehicle. The driver must still maintain situation awareness of the environment around the vehicle. ACC performs the safe following task normally requiring concentrated focal vision and distance estimation of a lead vehicle by the driver. LDW is a driver assistance system that alerts the driver when attempting a lane change without the use of turn signals or drifting inadvertently into another lane. The alert may be audio, visual, or tactile (vibration). The driver is required to intervene in-vehicle control as LDW does not take over control of the vehicle; it only provides an alert.

Additional ADAS have been developed, which extend assistance to the driver but still require driver involvement in the vehicle control process. These include forward control warning (FCW), lane-keeping assistance (LKA), and active steering (AS). FCW was designed to warn a driver of impending collision with other vehicles or objects. The FCW warning may be visual, audio or tactile, or some combination of these. FCW in its basic form is not a collision avoidance system but provides only a warning of a collision triggered by the rate of closure of the driver's vehicle on an object. AS is a system which assists the driver by adding additional torque to the steering wheel to aid in negotiating tight turns.

These and other ADAS systems can be operated singly or in combination for any given vehicle. Discussion is limited to the five systems mentioned as they are either already in operation in many vehicles or soon will be. Of particular concern here is the extent to which these systems, either singly or collectively, impact driving behavior, specifically the management and distribution of driver physical and mental resources. In general, alert and warning systems are intended to supplement driver attention which may, for whatever reason, be focused on other tasks whether driving-related or not. For most of these other tasks, driver focal visual attention will engage as driving is largely a visual task. This means that visual alerts will compete for attention with driver visual attention while non-visual alerts will not. Systems that provide actual assistance to the driver in performing a control task have a different and more substantial impact on driver resources. In the case of systems which impact vehicle control tasks such as LKA and ACC, driver visuomotor control resource requirements in both lateral and longitudinal axes are eliminated. While lane position and speed control are largely determined by pre-attentive, ambient vision in the visual periphery, control with respect to closure on a lead vehicle or other road object requires visual attention. Earlier versions of cruise control were not adaptive and required drivers to attend to lead vehicles and other hazards and overrule cruise control through braking or other actions.

In order to introduce some clarity with regard to resource management, the separation of ADAS systems into alerting and warning systems (e.g.,

LDW) on the one hand and partial automation systems (e.g., ACC) on the other will allow a more thorough examination of how these systems impact driver resources. ADAS systems examined here are only a partial list of many such systems in use or under development, but the impact of the systems discussed should provide an indication of how other systems will impact resource management.

ALERTING AND WARNING SYSTEMS

Among the first of the ADAS systems to be fielded was the LDW, designed to alert drivers about to move into an adjacent lane without signaling. The LDW warns the driver using visual, auditory, or tactile signals. The intent of the system is to alert the driver to the hazard of changing lanes without signaling while avoiding overwhelming the execution of other tasks which may then be disrupted. How these alerts impact resources employed in-vehicle tasks which may compete with LDW signal and the driver response to it is of interest as drivers are frequently engaged in non-driving tasks. As lane control is primarily of visual motor control tasks dependent on pre-attentive processes, tasks that compete for visual resources and manual control should affect, and be affected by, LDW events. Alternatively, secondary tasks which demand cognitive resources should not affect driving behavior in LDW events as the lane position control task is pre-attentive.

A field study of how LDW events would affect driver behavior in both single and dual-task conditions was conducted by McWilliams et al. (2016). The study compared the effects of different secondary tasks including the use of cell phones and information-entertainment (infotainment) systems on driver steering response and response to LDW events. The events were signaled by sound or tactile vibration. The secondary tasks required either visual-manual or auditory-vocal resources. Measured by steering wheel angle changes, secondary tasks which were visual-manual were more likely to result in greater SWA both before and after LDW events when compared to single task conditions and auditory-vocal secondary tasks. This indicates a resource competition between ongoing secondary, visual in-vehicle tasks and lane control which is visuomotor in nature.

Another study of LDW events and driver behavior in a closed-track instrumented vehicle assessed the impact of a cognitive secondary task on driver behavior (Cades et al., 2011). In this study, driver control behavior was investigated during LDW events with or without a mental math secondary task. The LDW event did not affect the driver's ability to perform the cognitive task nor did it affect variation in SWA.

In the longitudinal axis of the vehicle, the introduction of collision warnings such as the FCW was developed to alert the driver to an impending collision. It has evolved to provide automatic braking to these warning systems to provide additional vehicle protection. These systems, like other ADAS, are likely to be used when the driver is distracted. As a consequence, the impact of those distractions on the driver's response to the alert will be affected. The modality of the alert itself will also compete with perceptual resources used by the driver to carry out other tasks.

The importance of sensory modality resource utilization becomes apparent in the role of alerts for collision warnings. A number of studies have revealed that non-visual warnings for collision avoidance may be more effective than reliance on visual warnings. Non-visual warnings, auditory and vibrotactile, associated with the looming visual cue to collision discussed in an earlier chapter may appear to be more effective in attracting driver attention to the collision event (Gray, 2011; Ho et al., 2013; Lahmer et al., 2018). The non-visual looming cue increases in intensity as the looming stimulus (object) becomes closer. The advantage of non-visual warning is that it does not depend on focal visual attention to a display or in the case of manual driving, to the object itself. However, vibrotactile alerts, unlike auditory alerts, require that the driver is in continuous contact with some element of the vehicle such as the driving wheel or seat.

PARTIAL AUTOMATION

Increasingly, the automation of driving tasks has overtaken alerts and warnings in vehicle development. A few studies have examined how partial automation affects, and is affected by, distracting tasks. Automating driving tasks, unlike alerts and warnings, are not designed to attract driver attention, but to eliminate the need for the driver to focus resources on particular driving tasks. The desired effect of the design is to remove the task demand on the driver and thereby eliminating the need for drivers to expend related resources (perceptual, motor, and cognitive). The intended result is to improve driver safety by eliminating the driver, and therefore driver error, from the vehicle control loop. It is possible, however, the result may be that these same resources may simply be expended on increasing driver involvement with distracting tasks.

A review was recently conducted on the effects of automation on the influence of distracting tasks (Hungun et al., 2021). In total, 29 studies were reviewed, 24 of which involved vehicles equipped with ACC singly or in combination with LKA and AS. This combination automates control of the vehicle in both longitudinal and lateral vehicle axes. Automation of lane keeping (LKA) and following distance (ACC) resulted in a net increase in driver engagement with distracting tasks. These tasks included cell phone and vehicle infotainment system use, watching movie clips, reading and sending e-mail, eating, grooming, reading magazines, and a variety of verbal tasks involving anagrams and trivia quizzes. While these distracting tasks involved predominantly verbal-manual actions, they often required significant visual focal attention and cognitive processing as well. Generally, these distracting tasks concentrated driver attention within the vehicle rather than outside the vehicle where it is needed for control and collision avoidance. Drivers were also more likely to glance away from the roadway in support of the distracting tasks. Not surprisingly, performance on these distracting tasks improved in the presence of automation which likely provides the reinforcement for a driver to engage in more distractions. Overall, situation awareness of the road environment declined under automation.

The type of distracting task with which the driver engages has an effect on the ability of the driver to return to manual control should the need arise. Studies of the relative impact of distracting tasks on vehicle control with no automation present indicate that some of these tasks are likely to result in increasing the time needed to take over from the automated system. Cell phone use not only has the largest impact on manual control but was also found to be the most likely to be undertaken by drivers (Farah et al., 2016). Distracting tasks which generally have the greatest demand on driver manual control are more likely to affect the driver's ability to take over vehicle control from an automated system.

DETECTING FAILURES AND MANUAL TAKE OVER

Automated systems eventually require the driver to take over manual control either due to the driving situation exceeding the systems operational design domain (ODD) or to a system failure of one type or another. All automated systems are designed to operate within ODD, a domain described by limits in vehicle speed, traffic load, or other pre-determined factors. For more complex automated systems, the system will request the driver to take over from the automated system when the driving situation is approaching ODD limits. This take-over request (TOR) function is intended to allow the driver sufficient time to transition to manual control. The design of the TOR will involve visual, aural, or vibrotactile alerts, singly or in combination. Driver attention to distracting tasks will likely impair response to the TOR such that the driver may take too much time resulting in the system engaging emergency braking to avoid collision or road departure.

For the higher levels of automation such as SAE 3 or *conditional automation* even less demand on driver resources occurs. In SAE 3, the driver is not required to monitor current traffic conditions or other road environment events. This reduces dependence on driver resources normally required to monitor road conditions. The increasing emphasis on automated control and the consequent reduction of the driver to backup status have led to an interest in the types of non-driving, distracting tasks that the driver may engage in as a consequence. Of particular interest is how these distracting tasks affect the ability and time taken for drivers to take over manual control of driving tasks controlled by automated systems when needed. A review of distracting tasks used in studies of automated driving reflects the broad interest in this topic and the wide range of distracting tasks employed (Naujoks et al., 2018). A taxonomy of 21 different distracting tasks was identified in this study covering a broad range of driver resources. Once again, distracting tasks were predominantly those requiring visual focal attention as well as manual input. They range from the use of infotainment devices and navigation systems to a variety of tasks involving hand-held devices such as cell phones or tablets for playing games. Only a few of these tasks required speaking or listening, as occurs in phone use or watching movies, so auditory interference with the vehicle TOR has been the subject of only a few investigations when compared to visual-manual distracting tasks. This has implications for how automated systems interface with drivers specifically in the area of the take-over request alert function.

An extensive review of take-over time required of drivers in automated systems (partial and conditional) was conducted by Zhang et al. (2019) for the purpose of identifying driver response time and mitigating factors in take-over requests. The study covered 129 studies of takeovers of automated systems either due to critical situations or to actual TORs for the vehicle system. Take-over time measured from the TOR onset to the time the driver regained vehicle control averaged between 1.62 and 3.43 sec. However, as average take-over time correlated positively with the variation in that time, some take-over times averaged between 8 and 9 sec in length.

Average TORs by drivers in this study were affected by five factors. The first of these was the driver perception of urgency. Drivers tended to take longer to respond to TORs if they perceived that the criticality of the driving situation was not of a level demanding an immediate response. TORs were also affected by the presence of hand-held devices including cell phones. The driver interaction with these devices delayed responses to TORs significantly. Third, the higher level of automation (conditional vs. partial) resulted in a greater level of engagement with hand-held devices which resulted in a delayed response to the TOR. This increased level of engagement with the distracting task appears to result from the increased level of visual focal attention and cognitive processing resources which were made available by higher levels of automation. The fourth factor affecting TOR response times is the level of driver experience with TORs. This experience allowed the driver to identify critical conditions earlier as well as have practice with responding to TORs. Both of these factors reduced average TOR response times significantly. Finally, the sensory modality of TORs affected driver response times. Auditory or vibrotactile TORs resulted in faster take-over times than visual TORs likely due to the predominance of the resource demand of focal vision concentrated on visual-manual distractions with in-vehicle and hand-held devices.

Thus far, only incidents which generated a TOR from the automated system have been studied and then only under controlled conditions. A failure in the functioning of the automated system and the consequence for driving safety in uncontrolled, real-world driving is not yet known. In these real-life driving situations, inattentiveness of the driver in monitoring the performance of the system may result in take-over response times too slow to prevent collision or loss of control. Second, at this time in their development and deployment, the failure rate of these systems in actual use is unknown. Increasing reliance and trust in the operation of these systems by the driver which accrues with extended use will likely increase take-over times if an automated system were to fail. This is a result of a decline in driver attention to the operational state of a system believed to have high reliability.

Reviews of data on the effects of the automated system on driver behavior point to driver resources which, rather than increasing situation awareness of the vehicle state and road hazards, may simply increase driver engagement with distracting tasks. There is currently a strong association of distracting tasks with accidents in a driving environment in what are now predominantly manually controlled vehicles. The accident rate might actually increase with the addition of partial or conditional automation or higher levels simply because drivers are increasingly engaged in distracting tasks which the driver believes is warranted by their faith that the vehicle's automated systems will prevent collision or loss of control. If this is the result, then restrictions on automation which limit it to alerts and warnings or limited emergency control actions rather than the higher level of conditional automation may be necessary. Keeping the driver in the active control loop may be necessary to prevent a *psychological decoupling* from the driving task itself where the driver is no longer performing even the basic supervisory monitoring function. There is recent evidence that even a limited level of visual and manual vehicle interaction can be beneficial in keeping the driver in the vehicle control loop during conditional automation (Dillman et al., 2021). A limited form of automation assistance that keeps the driver in the control loop may be needed until the technology is fully developed to allow levels of full automation in which driver control in any form is no longer needed (SAE Level 5).

DRIVING ENVIRONMENT

Viewed as an integrated system of driver, vehicle, and roadway, the ability to manage driver resources is dependent on driver skill, vehicle design, and the design of road systems. The latter issue of road design needs to consider that the combined effects of automation and an aging driver population mean that a variety of design specifications may be impacted. Applying a resource management framework to the road environment is similar to how it might be applied to vehicle design. The emphasis should be on enhancing the efficiency at which driver resources, particularly ambient vision and focal vision, are expended by the driver. The second element of resource management is the need to increase the use of non-visual modalities, such as auditory and vibrotactile, to reduce the demand on driver visual processing of information. This also takes advantage of resources such as auditory focal attention which can filter out non-driving-related information (e.g., passenger communications) in favor of driving-related tasks, especially those requiring immediate action. The third element is to improve situation awareness by reducing the complexity of the road environment not only visually, but in the need for the driver to maintain multiple elements in working memory. This is particularly important for the older driver as well as young, inexperienced drivers who have difficulty maintaining situation awareness, particularly in complex traffic environments. These and other elements of driver resource management are intended to improve driver resource allocation and efficacy.

Street and Road Lighting

Illuminating roadways at night increases the effectiveness of driver vision in terms of both visual acuity and contrast sensitivity allowing drivers to improve the efficacy of both vehicle control and hazard detection. As noted in an earlier chapter (Chapter 2), even small increases in road illumination can have benefits. This is especially true in urban environments where the use of high-beam headlights cannot be used due to the blinding effect it has on drivers in approaching vehicles. By increasing the illumination of roads, not only are potential vehicle hazards detectable earlier, road signage will become more legible and road markings will have higher visual contrast particularly if retro-reflective paints are used.

When current roadway lighting is available, it is largely used as a means of deterring crime. The luminance output of streetlights is such that much of the roadway, bicycle lanes, and sidewalks are unlit. Hazards in these areas which do not have onboard lighting, such as pedestrians and non-powered vehicles, often are difficult to detect at a distance beyond the effective range of low-beam headlights. The greater the distance at which the hazard is detected, the more time drivers will have to respond with braking or maneuvering to avoid collision.

Intersections remain the most hazardous of roadway elements due to the increased likelihood of traffic and pedestrian hazards. They are particularly hazardous for older drivers and for distracted driver due to the problem of limits on visual focal attention capacity in complex traffic environments. Improvement in lighting can improve the efficacy of visual search and of visual focal attention in detecting and classifying hazards.

ROAD SIGNAGE

Daytime or photopic driving generally provides sufficient luminance values to allow the perception of the critical object detail in road signage. In the U.S., road signage legibility was established in 1947 by the American Medical Association based on a Snellen acuity of 20/40 (6/12 m). Two components of signage details are of importance. The first is the thickness of the detail, such as the strokes of numerals in a speed limit sign, and the second is the width of gaps in image detail as occurs between the strokes of the numerals. At the 20/40 acuity level, drivers should be able to discriminate details which have a stroke width and a gap between details of 2 min of arc or larger. By calculating the desired distance at which the sign needs to be legible, the actual physical stroke width and gap sizes can be calculated accordingly. Nominally, this standard is acceptable in most photopic conditions found in daytime driving. However, in lower light conditions where luminance levels fall below 400 cd/m² photopic vision is no longer optimal. This occurs in fog, in heavily forested areas, and in twilight or dusk conditions where drivers may not be using headlights to illuminate road signage.

An important ingredient in signage is the relative difference in luminance between the details of the signage, such as the numerals in a speed limit sign, and the background on which the details are placed. This luminance contrast allows the photoreceptors of the eyes to discriminate the details of an image from the background. Maximizing the contrast is often done using black lettering on a white background. A reflective white is used to further increase luminance contrast when the sign is struck by sunlight or by a vehicle's headlights. While signage colors such as a red background and white lettering for stop signs allow for improved sign identification, care should be taken in using colors for details. Signage that uses color contrast for segregating details from the background, such as orange on yellow background, will tend to have less luminance contrast and poorer legibility as a result.

Available lighting as discussed earlier will impact signage legibility. However, driver visual acuity also affects sign legibility. With the mesopic conditions of night driving, driver visual acuity declines. A study of healthy subjects' visual acuity showed a drop from an average of 20/17 to 20/66 when the light was reduced from photopic to mesopic levels (Hiraoka et al., 2015). The decline of acuity indicates that the effective legibility of a sign will be at a distance much less than is normally assumed. A driver with an effective visual acuity of 20/60, for example, will need to be much closer to a road sign to achieve legibility than a driver with the existing standard of 20/40.

The effects of reduced luminance levels at night suggest that signage legibility standards may need to be changed from the standard. Another factor is the increasing proportion of older drivers. As discussed in Chapter 6, older drivers' visual acuity begins to decline with increasing loss of acuity with increasing age. While the average older drivers' visual acuity does not drop below 20/40 until very old age, the variability or range of visual acuity increases with age. This means that increasingly larger numbers of older drivers will drop below the 20/40 limit with increasing age.

With an increasing proportion of older drivers on the horizon, consideration needs to be given to changing this standard. The higher legibility requirements would also address the reduced visual acuity of mesopic driving as well as some of the marginal photopic conditions mentioned above. A standard of a Snellen acuity of 20/60 (6/24 m),[1] for example, would provide better road signage legibility under a broader range of lighting conditions and driver visual acuity.

Modern signage does not need to be a static image on a board. The use of LEDS in signage allows not only high contrast for legibility in photopic and mesopic driving conditions but can be used for dynamic messaging as well. An additional utility of this signage is the provision of radar detection of oncoming vehicles and the display of their speed. The radar-generated speed display includes the feature of changing LED colors as the vehicle approaches and exceeds the speed limit. In speed-critical areas such as schools and more severe road curves, these features can be of great benefit in driver speed reduction. This is due in large part to the tendency of drivers to underestimate speed when not attending to the speedometer. Underestimation of speed is common, particularly in the lower speed range typical of urban environments.

ROADSIDE DISTRACTIONS

While in-vehicle devices and tasks dominate the discussion of driving distractions, distractions outside of the vehicle have long been an issue, particularly in urban environments. In the U.S., major interstate highways and freeways have largely eliminated roadside distractions such as advertising. However, advertising signage stills remain a problem in many rural and urban road environments. Advertising, by its very nature and design, is intended to attract driver attention away from the road and to the messages they display.

Limited evidence is available on the impact of advertising on driver visual focal attention and how this resource is shifted from its primary purpose of vehicle control and collision avoidance and toward the advertising signage. A review of advertising studies evaluating the effects of advertising signage on driver behavior was conducted by Wallace (2003). The review found evidence that advertising signage did indeed impact driver behavior, but the effects were situation specific. Advertising impacted driver visual search behavior on the roadway especially in urban environments by adding to visual clutter. The effect was to slow the visual processing needed for road hazard detection by incrementally adding to the visual search task. Advertising signage at critical points in road curvature interfered with visual focal attention to the tangent point of the curve necessary for steering angle adjustments to changes in road curvature. In a later study by Young et al. (2009), advertising was found to be detrimental to lateral vehicle control, to mental workload, and to visual attention. In some instances, the advertising signage diverted attention away from important road signage.

A more recent review not only found further evidence that advertising interfered with visual focal attention to the road environment, but that it also increased crash risk (Oviedo-Trespalacios et al., 2019). This was especially true for advertising signage that was electronically displayed and capable of frequent changes. The effects of advertising were particularly evident for younger drivers.

Additional research is needed to identify what specific design features of advertising signage are impacting driver behavior such as its placement with respect to driver visual search patterns on roadways. This may result in improved signage which has much less impact on driver behavior. However, if a determination is made that advertising signage is detrimental to driving safety, banning advertising signage entirely on rural and urban roadways may be required as is currently the case for major highways.

ROAD DESIGN

Eliminating visual distractions and improving lighting and signage design all contribute to the ability of drivers to manage resources, particularly in visual focal attention. Both efficiency and effectiveness of this resource, a resource needed to maintain vehicle control and avoid collision hazards, should be a primary aim in future road environment design.

The road itself can improve driver resource management of visual processes such as pre-attentive, and ambient visual processes involved in the perception of vehicle speed and collision avoidance. The former is derived by the driver from the visual perception of optic flow from the visual periphery. The latter is a normal reactive response to visual objects entering the driver's visual periphery from the road edge.

In a previous chapter (Chapter 2), the strong influence of the road environment in generating optic flow for the driver was illustrated under a variety of road conditions. In one case, transverse line markings extending from the road edge inward were employed. The intent and subsequent effect of these markings was to exaggerate the optic flow of the driver by increasing the number of perceptible road elements. These road markings were shown to be useful as means of slowing vehicle speed in the areas preceding speed-critical road sections such as severe curves and road intersections. Such markings could also be used on steeply sloped downhill sections as well as in approaches to tunnels.

An additional, though untested, use of road surface markings might include increasing the texture density of road surfaces through the use of either differentially reflective aggregates or surface markings. Increasing the perceived texture density of a road surface is a known aid in the perception of distance. As drivers tend to underestimate the perceived distance to a fixed point, the addition of surface texture could aid in improving the accuracy of distance perception, for example, in approaching an intersection. Such markings would also enhance optic flow and the resultant perception of speed.

Apart from potential visual cues, non-visual cues are already in use on road surfaces. Vibrotactile cues are now used to avoid off-road excursions by vehicles. The *rumble strip* is a track of grooves in the road surface adjacent and perpendicular to the road's outer edge. When in contact with moving tires, vibrations are sent through the vehicle frame accompanied by a rumbling sound, hence the name.

The rumble strip can be applied elsewhere on the road surface. As a part of the lane marking system, the rumble strip can be a non-visual addition to normal visual road markings. These would aid in providing a non-visual aid in alerting the driver to lane drift or departure similar to the concept of the lane departure warning used in partially automated vehicles. The rumble strip may also be used as a substitute for raised reflectors which are sometimes used to improve lane identification. However, these reflectors are known to interfere with sensors on automated vehicles and are being removed from some roads as a result.

ROAD INTERSECTIONS

A notable finding in driving behavior is the difficulty drivers, especially older drivers, experience when negotiating complex intersections. A complex intersection is characterized by a combination of high vehicle density, multiple traffic flow patterns, obliquely angled intersections, unprotected turns, and pedestrian traffic at multiple crossings. Two major driver resources, visual focal attention and situation awareness, are heavily tasked by complex intersections. Visual focal attention, supported by a disciplined visual search, needs to be moved around the intersection to identify and keep track of potential hazards. The task demand increases with traffic density and the presence of pedestrians. Situation awareness needs to be maintained by storing the current state of all potential hazards in working memory and updating these as they move in the intersection. The greater the number of potential hazards in the intersection, the greater the demand for both visual focal attention and working memory storage and retrieval.

Even relatively simple T-intersections, as are found in the intersection of minor to major roadways, can be problematic when traffic flow increases and pedestrian traffic is added. In a study by Werneke and Vollrath (2012), small increases in the flow of traffic on the major road coupled with the presence of pedestrians increased the demand on driver attention allocation significantly. A notable failure of allocation of attention was noted for drivers entering the major road from the minor road entrance. Drivers on minor roads typically, as in this study, enter a major roadway from an unsignalled intersection. This requires the driver to judge the distance and closure rate of oncoming traffic to the driver's left. The additional task of monitoring pedestrian crossing traffic on the driver's right placed further demand on driver visual attention. When traffic flow was increased and pedestrian traffic was added to the major roadway on the opposite side of the intersection, the demand for driver attention increased even further.

Road intersection design is driven by a variety of factors including traffic flow, pedestrian crossings, access to businesses and other properties, and the cost associated with traffic signaling. The need for improved safety, particularly at intersections, suggests that one of the key factors in design is a consideration for how driver visual focal attention is allocated and how the design impacts the driver's ability to maintain situation awareness. This problem is compounded by the increasing presence of older aged drivers in the population. These drivers are at the greatest risk of visual focal attention-related problems brought on by complex intersections and are more likely to be involved in accidents at these intersections than younger experienced drivers (Romoser et al., 2013; Dukic and Broberg, 2012).

For the purpose of evaluating the resources that a driver needs to allocate for a given intersection design, the designer needs to address the two resource issues mentioned previously. The first is visual focal attention defined as the duration of time needed to focus attention on a potential hazard whether hidden or not. Typically this is vehicle traffic including motorized and non-motorized but also includes pedestrian traffic. The number of fixations, the unit time per fixation, and the time between fixations or visual search time. Typical fixation times per hazard are about 0.3

to 0.5 sec for identification and classification. The time between fixations is generally driven by the distances between hazards with individual scan times of between 0.1 to less than 0.3 sec. The visual scanning and fixations within a given intersection will need to be repeated for those hazards that are in motion or are likely to be in motion. This allows for an aggregate of time to allocate visual attention in a given intersection. Traffic signaling, particularly turn signaling, are important variables in attention allocation at intersections as their presence minimizes the need for drivers to attend to potentially conflicting traffic.

SUMMARY

This chapter describes how resource management can be used as a framework for understanding driver behavior and for guidance in how vehicles and road systems can be designed. While there are other ways of viewing driver behavior which focus solely on driver attributes such as risk perception and driving capability, resource management as a framework focuses on the basic capabilities of drivers, physical and mental, and how they relate to vehicle and road design. Resource management is not only a function of driver resources but how they are employed through driver skill. Driver skill training is particularly important in the automatization of visuomotor control skills, in the allocation of visual focal attention, in task management, and in the perception of hazards. Vehicle design, especially automation, will play an increasing role in driving. Increasing automation may lead to a psychological decoupling of the driver from the driving task as seen in recent studies. This may lead to increasing levels of distraction and reduced driving safety. A variety of road design issues are noted which impact the limited resources available to drivers. Visual focal attention and situation awareness are particularly vulnerable in the complex intersection where attention to traffic and pedestrian behavior can exceed the resource capacity of even experienced drivers. Older drivers, an increasingly significant proportion of drivers, are especially vulnerable to road environments which require high levels of attention and situation awareness.

NOTE

1. At the time of writing, the state of California allows daytime waivers for drivers with 20/60 acuity.

References

Abdullah, A.S., Karim, M.R., Yamanaka, H., and Okushima, M. "Empirical analysis on the effect of gross vehicle weight and vehicle size on speed in car following situation." *Asian Transport Studies* 2, (2013): 351–362.tw

Ahmadi, N., Romoser, M., and Katrahmadi, A. "Short and long term transfer of training in tablet-based teen driver hazard perception training program." *Proceedings of the Human Factors and Ergonomics Society September* (2018). https://doi.org/10.1177/1541931218621445.

Alvarez, F.J. and Fierro, I. "Older drivers, medical condition, medical impairment and crash risk." *Accident Analysis & Prevention* 40, (2008): 55–60.

Andersen, G.J. and Sauer, C.W. "Optical information for car following: The driving by visual angle (DVA) model." *Human Factors* 49, (2007): 878–896.

Andrea, D.J., Fildes, B.N., and Triggs, T.J. "The sensitivity and bias of older and younger driver judgements in complex traffic environments." In *Road Safety Research, Policing & Enforcement Conference*, Melbourne, Australia. November, 2001.

Anson, E. and Jeka, J. "Perspectives on aging vestibular function." *Frontiers in Neurology* 6, (2016). https://doi.org/10.3389/fneur.2015.00269.

Anstey, K.J. and Wood, J. "Chronological age and age-related cognitive deficits are associated with an increase in multiple types of driving errors in late life." *Neuropsychology* 25, (2011): 613–621.

Atchley, P. and Andersen, G.J. "The effect of age, retinal eccentricity, and speed on the detection of optic flow components." *Psychology and Aging* 13, (1998): 297–308.

Ball, K., Owsley, C., Sloane, M.E., Roenker, D.L., and Bruni, J.R. "Visual attention problems as a predictor of vehicle crashes in older drivers." *Investigative Ophthalmology & Visual Science* 34, (1993): 3110–3123.

Beanland, V. and Wynne, R.A. "Does familiarity breed competence? Effects of driver experience, road type and familiarity on hazard perception." *Proceedings of the Human Factors and Ergonomics Society 2019 Annual Meeting*, Seattle, WA. October, 2019: 2006–2010.

Bian, Z. and Andersen, G.J. "Aging and the perception of egocentric distance." *Psychology and Aging* 28, (2013): 813–825.

Bian, Z., Kang, J.J., and Andersen, G.J. "Changes in extent of spatial attention with increased workload in dual-task driving." *Transportation Research Record* 2185, (2010): 8–14.

Billino, J., Bremmer, F., and Gegenfurtner, K.R. "Differential aging of motion processing mechanisms: Evidence against general perceptual decline." *Vision Research* 48, (2008): 1254–1261.

Billino, J. and Pilz, K.S. "Motion perception as a model for perceptual aging." *Journal of Vision* 19, (2019): 1–28.

Bolstad, C.A. "Situation awareness; Does it change with age?" *Proceedings of the Human Factors and Ergonomics Society Annual Meeting* 45, (2001): 272–276.

Borhan, N., Ibrahim, M.K.A., and Ab Rashid, A.A. "Hazard detection among young and experienced drivers via driving simulator." *Journal of the Society of Automotive Engineers Malaysia* 3, (2019): 20–31.

Borowsky, A., Oron-Gilad, T., and Parmet, Y. "Age and skill differences in classifying hazardous traffic scenes." *Transportation Research Part F: Traffic Psychology and Behaviour* 12, (2009): 277–287.

Borowsky, A., Oron-Gilad, T., and Parmet, Y. "The role of driving experience in hazard perception and categorization: A traffic-scene paradigm." *International Journal of Social, Behavioral, Educational, Economic, Business and Industrial Engineering* 4, (2010): 305–309.

Bowers, A., Peli, E., Elgin J., McGwin Jr. G.E., Owsley, C. "On-road driving with moderate visual field loss." *Optometry and Vision Science* 82, (2005): 657–667.

Brookhuis, K.A., de Vries, G., and de Waard, D. "The effects of mobile telephoning on driving performance." *Accident Analysis and Prevention* 23, (1991): 309–316.

Brown, I.D., Tickner, A.H., and Simmonds, D.C.V. "Interference between concurrent tasks of driving and telephoning." *Journal of Applied Psychology* 53, (1969): 419–424.

Burney, G.M. "Estimation of distances while driving." (No. NSR 262 Monograph), 1977.

Cain, M.S., Adamo, S.H., and Mitroff, S.R. "A taxonomy of errors in multiple-target visual search." *Visual Cognition* 21, (2013): 899–921.

Caird, J.K. and Hancock, P.A. "The perception of arrival time for different oncoming vehicles at an intersection." *Ecological Psychology* 6, (1994): 83–109.

Caird, J.K., Willness, C.R., Steel, P., and Scialfa, C. "A meta-analysis of the effects of cell phones on driver performance." *Accident Analysis & Prevention* 40, (2008): 1282–1293.

Calvi, A. "Does roadside vegetation affect driving performance?" *Transportation Research Record: Journal of the Transportation Research Board, No. 2518*, (2015): 1–8.

Cao, S., Samuel, S., Muzello, Y., Ding, W., Zahang, X., and Niu, J. "Hazard perception in driving: A systematic literature review." *Transportation Research Record: Journal of the Transportation Research Board* (June, 2022). https//doi.org/10.1177/036119119812210966.

Castro, C., Martínez, C., Tornay, F.J., Fernández, P.G., and Martos, F.J. "Vehicle distance estimations in nighttime driving: A real-setting study." *Transportation Research Part F: Traffic Psychology and Behaviour* 8, (2005): 31–45.

Cavallo, V. and Laurent, M. "Visual information and skill level in time-to-collision estimation." *Perception* 17, (1988): 623–632.

Cavallo, V., Colomb, M., and Dore, J. "Distance perception of vehicle rear lights in fog." *Human Factors* (September 2001). https://doi.org/10.1518/001872001775898197.

Chapman, J.R. and Noyce, D.A. "Influence of roadway geometric elements on driver behavior when overtaking bicycles on rural roads." *Journal of Traffic and Transportation Engineering* (English Edition) 1, (2014): 28–38.

Chapman, P.R. and Underwood, G. "Visual search of driving situations: Danger and experience." *Perception* 27, (1998): 951–964.

Charlton, S.G. "The role of attention in horizontal curves: A comparison of advance warning, delineation, and road marking treatments." *Accident Analysis and Prevention* 39, (2007): 873–885.

Chattington, M., Wilson, M., Ashford, D., and Marple-Horvat, D.E. "Eye-steering coordination in natural driving." *Experimental Brain Research* 180, (2007): 1–14.

Chatziastros, A., Wallis, G.M., and Bulthoff, H. "The effect of field of view and surface texture in driving steering performance." In A.G.Cole, I.D. Brown, C.M. Haslegrave, and S.P. Taylor (Eds.) *Vision in vehicles VII (Amsterdam).* North-Holland: Elsevier Science B.V., 1999.

Chevalier, A., Coxon, K., Chevalier, A.J., Clarke, E., Rogers, K., Brown, J., Boufous, S., Ivers, R., and Keay, L. "Predictors of older drivers' involvement in rapid deceleration events." *Accident Analysis & Prevention* 98, (2017): 312–319.

Chrysler, S.T., Danielson, S.M., and Kirby, V.M. "Age differences in visual abilities in nighttime driving field conditions." *Proceedings of the Human Factors and Ergonomics Society Annual Meeting* 40, (1996): 923–927. Sage, CA: SAGE Publications.

Clark, B. and Stewart, J.D. "Effects of angular acceleration on man: Thresholds for the perception of rotation and the oculogyral illusion." *Aerospace Medicine* 40, (1969): 952–956.

Clay, O.J., Wadley, V.G., Edwards, J., Roth, D.L., Roenker, D., and Ball, K.K. "The useful field of view as a predictor of driving performance in older adults: A cumulative metaanalysis." *Optometry & Vision Science* 82, (2005): 724–731.

Colbourn, C.J., Brown, I.D., and Copeman, A.K. "Drivers' judgment of safe distances in vehicle following." *Human Factors* 78, (1978): 1–11.

Comte, S.L. and Jamson, A.H. "Traditional and innovative speed reducing measures for curves: An investigation of driver behavior using a driver simulator." *Safety Science* 36, (2000): 137–150.

Coutton-Jean, C., Mestre, D.R., Goulon, C., and Bootsma, R.J. "The role of edge lines in curve driving." *Transportation Research Part F* 12, (2009): 483–493.

Cox, J.A., Beanland, V., and Filtness, A.J. "Road and safety perception on urban and rural roads: Effects of environmental features, driver age and risk sensitivity." *Traffic Injury Prevention* 18, (2017): 703–710.

Crawford, A. "The overtaking driver." *Ergonomics* 6, (1963): 153–170.

Crick, J. and McKenna, F.P. "Hazard perception can it be trained?" *Behavioural Research in Road Safety II, Proceedings of a Seminar, 17–18 September, 1992,* Manchester University, Manchester, UK.

Crundall, D. "Hazard prediction discriminates between novice and experienced drivers." *Accident Analysis and Prevention* 86, (2016): 47–48.

Crundall, D., Shenton, C., and Underwood, G. "Eye movements during intentional car following." *Perception* 33, (2004): 975–986.

Cutting, J.E. and Vishton, P.M. "Perceiving layout and knowing distances: The integration, relative potency, and contextual use of different information about depth." In *Perception of Space and Motion*, 69–117. New York: Academic Press, 1995.

Dastrup, E., Lees, B.N., Dawson, J.D., Lee, J.D., and Rizzo, M. "Differences in simulated car following behavior of younger and older drivers." In *Driving Assessment Conference* (Vol. 5, No. 2009). University of Iowa.

Deery, H.A. "Hazard and risk perception among young novice drivers." *Journal of Safety Research* 30, (1999): 225–236.

DeLucia, P., Kaiser, M., Bush, J., Meyer, L., and Sweet, B. "Information integration in judgements of time to contact." *The Quarterly Journal of Experimental Psychology Section A* 56, (2003): 1165–1189.

DeLucia, P.R. and Tharanathan, A. "Responses to deceleration during car following: Roles of optic flow, warnings, expectations, and interruptions." *Journal of Experimental Psychology: Applied* 15, (2009): 334–350.

de Waard, D., Steyvers, F.J.J.M., and Brookhuis, K.A. "How much visual road information is needed to drive safely and comfortably?" *Safety Science* 42, (2004): 639–655.

de Waard, D., Chris Dijksterhuis, C., and Karel A. Brookhuis, K.A. "Merging into heavy motorway traffic by young and elderly drivers." *Accident Analysis & Prevention* 41, (2009): 588–597.

Dillmann, J., den Hartigh, R.J.R., Kurpiers, C.M., Raisch, F.K., de Waard, D., and Cox, R.F.A. "Keeping the driver in the loop in conditionally automated driving: A perception-action theory approach." *Transportation Research Part F* 79, (2021): 49–620.

Donges, E. "A two-level model of driver steering behavior." *Human Factors* 20, (1978): 691–707.

Doshi, A. and Trivedi, M.M. "On the roles of eye gaze and head dynamics in predicting driver's intent to change lanes." *IEEE Transactions on Intelligent Systems* 10, (2009): 453–462.

Drews, F.A., Paupathi, M., and Strayer, D.L. "Passenger and cell phone conversations in simulated driving." *Journal of Experimental Psychology: Applied* 14, (2008): 392–400.

Drews, F.A., Yazdani, H., Godfrey, C.N., Cooper, J.M., and Strayer, D.L. "Text messaging during simulated driving." *Human Factors* 51, (2009): 762–770.

Dukes, J., Norman, J.F., Shapiro, H., and Peterson, A. "Aging and outdoor visual distance perception." *Journal of Vision* 20, (2020): 136–136.

Dukic, T. and Broberg, T. "Older drivers' visual search behaviour at intersections." *Transportation Research Part F: Traffic Psychology and Behaviour* 15, (2012): 462–470.

Duncan, J., Williams, P., Nimmo-Smith, I., and Brown, I.D. "The control of skilled behaviour: Learning, intelligence and distraction." In D.E. Meyer and S. Kornblum (Eds.) *Attention and Performance XIV*. MIT Press: Cambridge, MA, 1992.

Durkee, S. and Ward, N. "Effect of driving simulation parameters related to egomotion on speed perception." In *Driving Assessment Conference* 6, 2011. University of Iowa.

Easa, S.M., Reed, M.J., Russo, F., Dabbour, E., Mehmood, A., and Curtis, K. "Effect of increasing road Light Luminance on night driving performance of older adults." *World Academy of Science, Engineering and Technology* 44, (2010): 325–332.

Edquist, J., Rudin-Brown, C.M., and Lenné, M.G. "The effects of on-street parking and road environment visual complexity on travel speed and reaction time." *Accident Analysis & Prevention* 45, (2012): 759–765.

Endsley, M.R. "Measurement of situation awareness in dynamic systems." *Human Factors* 37, (1995): 65–84.

Eriksson, L., Palmqvist, L., Hultgren, J.A., Blissing, B., and Nordin, S. "Performance and presence with head-movement produced motion parallax in simulated driving." *Transportation Research Part F: Traffic Psychology and Behaviour* 34, (2015): 54–64.

Evans, D.W. and Ginsburg, A.P. "Contrast sensitivity predicts age-related differences in highway-sign discriminability." *Human Factors* 27, (1985): 637–642.

Farah, H., Zatmeh, S., Toledo, T., and Wagner, P. "Impact of distracting and drivers' cognitive failures on driving performance." *Advances in Transportation Studies* Special Issue 1, (2016): 71–28.

Farley, N.J., Norman, H.F., Craft, A.E., Walton, C.L., Bartholomew, A.N., Burton, C.L., Wiesemann, E.Y., and Crabtree, C.E. "Stereopsis and aging." *Vision Research* 48, (2008): 2456–2465.

Figueira, A.C. and Larocca, A.P.C. "Analysis of the factors influencing overtaking in two-lane highways: A driving simulator study." *Transportation Research Part F: Psychology and Behaviour* 69, (2020): 38–48.

Fildes, B.N. and Triggs, T.J. "The effect of changes in curve geometry on magnitude estimates of road-like perspective geometry." *Perception and Psychophysics* 37, (1985): 318–224.

Fitch, G.M., Blanco, M., Morgan, J.F., and Wharton, A.E. "Driver braking performance to surprise and expected events." In *Proceedings of the Human Factors and Ergonomics Society Annual Meeting* (Vol. 54, No. 24, pp. 2075–2080). Sage, CA: Los Angeles: SAGE Publications, 2010.

Franconeri, S.L. and Simons, D.J. "Moving and looming stimuli capture attention." *Perception and Psychophysics* 65, (2003): 999–1010.

Frissen, I. and Mars, F. "The effect of visual degradation on anticipatory and compensatory control." *The Quarterly Journal of Experimental Psychology* 67, (2014): 499–507.

Fuller, R. "Towards a general theory of driver behavior." *Accident Analysis and Prevention* 37, (2005): 461–472.

Fuller, R. "Driver control theory: From task difficulty homeostasis to risk allostasis." In B.E. Porter (Ed.) *Handbook of Traffic Psychology*. North-Holland: Elsevier, 2011.

Gegenfurtner, K.R., Mayser, H., and Sharpe, L.T. "Seeing movement in the dark." *Nature* 398, (1999): 475–476.

Godley, S.T., Triggs, T.J., and Fildes, B.N. "Speed reduction mechanisms of transverse lines." *Transportation Human Factors* 2, (2000): 97–312.

Godley, S.T., Triggs, T.J., and Fildes, B.N. "Perceptual lane width, wide perceptual road centre markings and driving speeds." *Ergonomics* 47, (2004): 237–256.

Gould, M., Poulter, D.R., Helman, S., and Wann, J.P. "Judgments of approach speed for motorcycles across different lighting levels and the effect of an improved tri-headlight configuration." *Accident Analysis & Prevention* 48, (2012): 341–345.

Gray, R. "Looming auditory warnings for driving." *Human Factors* 53, (2011): 63–74.

Gray, R. and Reagan, D. "Accuracy of estimating time to collision using binocular and monocular information." *Vision Research* 38, (1998): 499–512.

Gray, R. and Regan, D.M. "Perceptual processes used by drivers during overtaking in a driving simulator." *Human Factors* (June, 2005). https://doi.org/10.1518720054679443.

Green, M. "How long does it take to stop? Methodological analysis of driver perception-brake times." *Transportation Human Factors* 2, (2000): 195–216.

Gruber, N., Mosimann, U.P., Müri, R.M., and Nef, T. "Vision and night driving abilities of elderly drivers." *TrafficInjury Prevention* 14, (2013): 477–485.

Gugerty, L. "Situation awareness in driving." In D.L. Fisher, M. Rizzo, J. Caird, and J.D. Lee (Eds.) *Handbook for Driving Simulation in Engineering, Medicine and Psychology.* Boca Raton, FL: CRC Press, 2011.

Haegerstrom-Portnoy, G, Schneck, M.E., and Brabyn, J.A. "Seeing into old age: Vision function beyond acuity." *Optometry and Vision Science* 76, (1999): 141–158.

Haibach, P., Slobounov, S., and Newell, K. "Egomotion and vection in young and elderly adults." *Gerontology* 55, (2009): 637–643.

Hakamies-Blomqvist, L., Raitanen, T., and O'Neill, D. "Driver ageing does not cause higher accident rates per km." *Transportation Research Part F: Traffic Psychology and Behaviour* 5, (2002): 271–274.

Hanson, T.R. and Hildebrand, E.D. "Are rural older drivers subject to low-mileage bias?" *Accident Analysis & Prevention* 43, (2011): 1872–1877.

He, J., Chaparro, A., Nguyen, B., Burge, R.J., Crandall, J., Chaparro, B., Ni, R., and Cao, S. "Texting while driving: Is speech-based text entry less risky than handheld text entry?" *Accident Analysis and Prevention* 72, (2014): 287–295.

Herbert, N.C., Thyer, N.J., Isherwood, S.J., and Merat, N. "The effect of auditory distraction on the useful field of view in hearing impaired individuals and its implications for driving." *Cognition, Technology & Work* 18, (2016): 393–402.

Hickson, L., Wood, J., Chaparro, A., Lacherez, P., and Marszalek, R. "Hearing impairment affects older people's ability to drive in the presence of distracters." *Journal of the American Geriatrics Society* 58, (2010): 1097–1103.

Hiraoka, T., Hoshi, S., Okamoto, Y., Okamotot, F., and Tetsuro, O. "Mesopic functional visual acuity in normal subjects." *PLoS One* 10, (2015): e0134505.

Hirsch, P., Bellavance, F., Tahari, S., and Faubert, J. "Towards the validation of a driving simulator-based hazard response test for novice drivers." *Proceedings of the Eighth International Driving Symposium on Human Factors in Driving Assessment, Training, and Vehicle Design* (2015): 338–344.

Ho, C., Spence, C., and Gray, R. "Looming auditory and vibrotactile collision warnings for safe driving." In *Proceedings of the Seventh International*

Driving Symposium on Human Factors in Driver Assessment, Training, and Vehicle Design 7, (2013).

Horberry, T., Andersen, J., and Regan, M. "The possible safety benefits of enhanced road markings: A driving simulator evaluation." *Transportation Research Part F Traffic Psychology and Behavior* 9, (2006): 77–87.

Horrey, W.J. and Wickens, C.D. "Multiple resource modeling of task interference in vehicle control, hazard awareness and in-vehicle task performance." *Proceedings of the Second International Driving Symposium on Human Factors in Driver Assessment, Training, and Vehicle Design* 2, (2003): 7–12.

Horrey, W.J. and Wickens, C.D. "Examining the impact of cell phone conversations on driving using meta-analytic techniques." *Human Factors* 48, (2006): 196–205.

Horswill, M.S., Anstey, K.J., Hatherly, C.G., and Wood, J.M. "The crash involvement of older drivers is associated with their hazard perception latencies." *Journal of the International Neuropsychological Society* 16, (2010): 939–944.

Horswill, M.S., Falconer, E.K., Pachana, N.A., Wetton, M., and Hill, A. "The longer-term effects a brief hazard perception training intervention in older adults." *Psychology and Aging* 30, (2015): 62–67.

Horswill, M.S., Helma, S., Ardiles, P., and Wann, J.P. "Motorcycle accident risk could be inflated by a time to arrival illusion." *Optometry and Vision Science* 82, (2005): 740–746.

Horswill, M.S., Taylor, K., Newnam, S., Wetton, M., and Hill, A. "Even highly experienced drivers benefit from a brief hazard perception training intervention." *Accident Analysis and Prevention* 52, (2013): 100–110.

Hungund, A.P., Pai, G., and Pradhan, A.K. "Systematic review of research on driver distraction in the context of advanced driver assistance systems." *Transportation Research Record* 2675,(2021): 1–10.

Ishigami, Y. and Klein, R.M. "Is a hands-free phone safer than a handheld phone?" *Journal of Safety Research* 40, (2009): 157–164.

Ivers, R., Senserrick, T., Boufous, S., Stevenson, M., Chen, H.-Y., Woodward, M., and Norton, R. "Novice drivers' risky behavior, risk perception, and crash risk: Findings from the DRIVE study." *American Journal of Public Health* 99, (September, 2009): 1638–1644.

Janssen, C.P. and Brumby, D.P. "Strategic adaptation to performance objectives in a dual-task setting." *Cognitive Science* 34, (2010): 1548–1560.

Janssen, C.P., Brumby, D.P., and Garnett, R. "Natural break points: The influence of priorities and cognitive and motor cues on dual-task interleaving." *Journal of Cognitive Engineering and Decision Making* 6, (2012): 5–29.

Johnson, C.A. and Keltner, J.L. "Incidence of visual field loss in 20,000 eyes and its relationship to driving performance." *Arch Ophthalmology* 101, (1983): 371–375.

Kaber, D., Zhang, Y., Jin, S., Mosaly, P., and Garner, M. "Effects of hazard exposure and roadway complexity on young and older driver situation awareness and performance." *Transportation Research Part F: Traffic psychology and Behaviour* 15, (2012): 600–611.

Kahana-Levy, N., Sahvitsky-Golkin, S., Borowsky, A., and Vakil, E. "The effects of repetitive presentation of specific hazards on eye movements in hazard perception training, of experienced and young-inexperienced drivers." *Accident Analysis and Prevention* 122, (2019): 255–267.

Kahneman, D. *Attention and Effort.* Englewood Cliffs, NJ: Prentice-Hall, 1973.

Kang, J.J., Ni, R., and Andersen, G.J. "Effects of reduced visibility from fog on car-following performance." *Transportation Research Record: Journal of the Transportation Research Board* 2069, (2008): 9–15.

Kasneci, E., Kasneci, G., Kuhler, T.C., and Wolfgang, R. "Online recognition of fixations, saccades, smooth pursuits for automated analysis of traffic hazard perception." In P. Koprinkova-Hristova, V. Mladenov, and N. Kasabov (Eds.) *Artificial Neural Networks.* Springer Series in Bio-Neuroinformatics 4, (2015): 411–434.

Kelly, J.W., Beall, A.C., Loomis, J.M., Smith, R.S., and Macuga, K.L. "Simultaneous measurement of steering performance and perceived heading on a curving path." *ACM Transactions on Applied Perception (TAP)*, 3(2006): 83–94.

Keltner, J.L. and Johnson, C.A. "Mass visual field screening in a driving population." *Ophthalmology* 87, (1980): 785–792.

Kennedy, R.S., Jentsch, F., and Al-Awar Smither, J. "Looming detection among drivers of different ages." *Proceedings of the Human Factors and Ergonomics Society Annual Meeting* 45, (2001). Sage, CA: SAGE Publications.

Kingma, H. "Thresholds for perception of direction of linear acceleration as a possible evaluation of the otolith function." *BMC Ear, Nose and Throat Disorders* 5, (2005): 1–6.

Klein, G. "Naturalistic decision making." *Human Factors* 50, (2008): 456–460.

Kline, D.W., Kline, T.J., Fozard, J.L., Kosnik, W., Schieber, F., and Sekuler, R. "Vision, aging, and driving: The problems of older drivers." *Journal of Gerontology* 47, (1992): 27–34.

Kokubun, M., Konishi, H., Kazumori, H., Kurashi, T., and Umemura, Y. "Assessment of drivers' risk perception using a simulator." *R&D Review of Toyota CDRL* 39, (2004): 9–15.

Kountourotis, G.K., Mole, C.D., Merat, N., and Wilkie, R.M. "The need for speed: Global optic flow speed influences steering." *Royal Society Open Science* 3, (May, 2016): 160096.

Kusano, K.D. and Gabler, H. "Method for estimating time to collision at braking in real-world, lead vehicle stopped rear-end crashes for use in pre-crash system design." *SAE International Journal of Passengers Cars – Mechanical Systems* 4, (2011): 435–443.

Lahmer, M., Glatz, C., Seibold, V.C., and Chuang, L.L. "Looming auditory collision warnings for semi-automated driving: An ERP study." *Proceedings of the 10th international conference on automotive user interfaces and interactive vehicular applications*, (2018): 310–319.

Land, M.F. "The visual control of steering." *Vision and Action* 28, (1998): 168–180.

Land, M.F. "Does steering a car involve perception of the velocity flow field?" In J.M. Zanker and J. Zeil (Eds.) *Motion Vision: Computational, Neural, and Ecological Constraints*, 227–235. Berlin: Springer Verlag, 2001.

Land, M.F. and Lee, D.N. "Where we look when we steer." *Nature* 369, (1994): 742–743.

Land, M. and Horwood, J. "The relations between head and eye movements during driving." *Vision in Vehicles* 5, (1996): 153–160.

Langford, J., Charlton, J.L., Koppel, S., Myers, A., Tuokko, H., Marshall, S., Man-Son-Hing, M., Darzins, P., Di Stefano, M., and Macdonald, W. "Findings from the Candrive/Ozcandrive study: Low mileage older drivers, crash risk and reduced fitness to drive." *Accident Analysis & Prevention* 61, (2013): 304–310.

Langford, J., Methorst, R., and Hakamies-Blomqvist, L. "Older drivers do not have a high crash risk—A replication of low mileage bias." *Accident Analysis & Prevention* 38, (2006): 574–578.

Larish, J.F. and Flach, J.M. "Sources of optical information useful for perception of speed of rectilinear self-motion." *Journal of Experimental Psychology: Human Perception and Performance* 16, (1990): 295–302.

Laurent, G. and Gabbiani, F. "Collision-avoidance: Nature's many solutions." *Nature Neuroscience* 1, (1998): 261–263.

Lee, A.T. *Vehicle Simulation: Perceptual Fidelity in Virtual Environments.* Boca Raton, FL: CRC Press, 2017.

Lee, D.N. "A theory of visual control of braking based on information about time-to-collision." *Perception* 5, (1976): 437–459.

Lee, D.N. and Lishman, R. "Visual control of locomotion." *Scandinavian Journal of Psychology* 18, (1977): 224–230.

Lee, S. and Koo, N. "Change of stereoacuity with aging in normal eyes." *Korean Journal of Ophthalmology* 19, (2005): 136–139.

Lee, Y. "Analysis of unintended acceleration through physical interference of accelerator." *Forensic Science International Reports* 2, (2020): 100079.

Lerner, N.D. "Brake perception-reaction times of older and younger drivers." *Proceedings of the Human Factors and Ergonomics Society Annual Meeting* 37, (1993): 206–210. Sage, CA: SAGE Publications.

Levulis, S.J., DeLucia, P.R., and Jupe, J. "Effects of oncoming vehicle size on overtaking judgments." *Accident Analysis and Prevention* 82, (2015): 163–170.

Levulis, S.J., DeLucia, P.R., Yang, J., and Nelson, V. "Does perceived harm underlie effects of vehicle size on overtaking judgments during driving?" *Proceedings of the Human Factors and Ergonomics Society Annual Meeting* 62, (2018): 1384–1388. Sage, CA: SAGE Publications.

Levy, J. and Pashler, H. "Task prioritization in multitasking during driving: Opportunity to abort a concurrent task does not insulate from dual-task slowing." *Applied Cognitive Psychology* 22, (2008): 507–525.

Lewis-Evans, B. and Charlton, S.G. "Explicit and implicit processes in behavioural adaptation to road width." *Accident Analysis and Prevention* 38, (2006): 610–617.

Li, L. and Chen, J. "Relative contribution of optic flow, bearing, and splay angle information to lane keeping." *Journal of Vision* 10, (2010): 1–14.

Liebermann, D.G., Ben-David, G., Schweitzer, N., Apter, Y., and Parush, A. "A field study on braking responses during driving. I. Triggering and modulation." *Ergonomics* 38, (1995): 1894–1902.

Liu, A. "What the driver's eye tells the car's brain." In G. Underwood (Ed.) *Eye Guidance in Reading and Scene Perception*, 431–452. Amsterdam: Elsevier Science Ltd., 1998.

Mackenzie, A.K. and Harris, J.M. "Eye movements and hazard perception in active and passive driving." *Visual Cognition* 23, (2015): 736–757.

Maddox, M.E. and Kiefer, A. "Looming threshold limits and their use in forensic practice" *Proceedings of the Human Factors and Ergonomics Society Annual Meeting* 56, (October, 2012): 700–704.

Markkula, G. "Modeling driver control behavior in both routine and near-accident driving." *Proceedings of the Human Factors and Ergonomics Society Annual Meeting* 58, (2014): 879–883. Sage, CA: SAGE Publications.

Markulla, G., Engstrom, J., Lodin, J., Bargman, J., and Victor, T. "A farewell to brake response in naturalistic rear-end emergencies? Kinematics-dependent brake response in naturalistic rear-end emergencies." *Accident Analysis & Prevention (Part A)* 95, (2016): 209–226.

Mayhew, D.R., Simpson, H.M., and Ferguson, S.A. "Collisions involving senior drivers: High-risk conditions and locations." *Traffic Injury Prevention* 7, (2006): 117–124.

McGhee, D.V., Roe, C.A., Boyle, L.N., Wu, Y., Ebe, K., Foley, J., and Angell, L. "The wagging foot of uncertainty." *SAE International Journal of Transportation Safety* 4, (2016): 289–294.

McDonald, C.C., Goodwin, M.A., Pradhan, A.K., Romoser, M.R.E., and Williams, A.F. "A review of hazard anticipation training programs for young drivers." *Journal of Adolescent Health* 57, (2015): S15–S23.

McKenna, F. and Crick, J. "Experience and expertise in hazard perception." In *Behavioural Research in Road Safety. Proceedings of a Seminar Held at Nottingham University*, 26–27 September 1990 (No. PA 2038/91). 1991.

McKenna, F.P. and Farrand, P. "The role of automaticity in driving." In G.B. Grayson (Ed.) *Behavioural Research in Road Safety IX, (20–25)*. Crowthorne, UK: Transport Research Laboratory, 1999.

McLeod, R.W. and Ross, H.E. "Optic flow and cognitive factors in time-to-collision estimates." *Perception* 12, (1983): 417–423.

McWilliams, T., Brown, D., Reimer, B., Mehler, B., and Dobres, J. "Observed differences in lane departure warning responses during single-task and dual-task driving: A secondary analysis of field driving data." SAE Technical Paper 2016-01-1425, 2016.

Merat, N. and Jamson, H. "A driving simulator study to examine the role of vehicle acoustics on drivers' speed perception." *Proceedings of the International Driving Symposium on Human Factors in Driver Assessment, Training and Vehicle Design* (June, 2011). University of Iowa: Iowa City, IA.

Milleville-Pennel, I., Jean-Michel, H., and Elise, J. "The use of hazard road signs to improve the perception of severe bends." *Accident Analysis and Prevention* 39, (2007): 721–730.

Monaco, W.A., Kalb, J.T., and Johnson, C.A. "Motion detection in the far peripheral visual field." *U.S. Army Research Laboratory, ARL-MR-0684* (December, 2007).

Moran, C., Bennet, J.M., and Prabhakharan, P. "Road user hazard perception tests: A systematic review of current methodologies." *Accident Analysis and Prevention* 129, (2019): 309–333.

Mole, C.D., Kountouriotis, G., Billington, J., and Wilkie, R.M. "Optic flow speed modulates guidance level control: New insights into two-level steering."

Journal of Experimental Psychology: Human Perception and Performance 42, (2016): 1818.

Muhlenen, A. and Lleras, A. "No-onset looming motion guides spatial attention." *Journal of Experimental Psychology* 33, (2007): 1297–1310.

Naatanen, R. and Summala, H. "A model for the role of motivational factors in drivers' decision-making." *Accident Analysis and Prevention* 6, (1974): 243–261.

Naujoks, F., Befeelein, D., Wiedemann, K., and Neukun, A. "A review of non-driving-related tasks used in studies on automated driving." In N. Stanton (Eds.) *Advances in Human Aspects of Transportation.* AFHE 2017, Advances in Intelligent Systems and Computer. Cham, Switzerland, 2018.

Ni, R., Kang, J., and Andersen, G.J. "Age-related driving performance: Effect of fog under dual task conditions." *Proceedings of the Fourth International Driving Symposium on Human Factors in Driver Assessment, Training, and Vehicle Design* (July, 2007): 365–370.

Norman, J.F., Adkins, O.C., Dowell, C.J., Shain, L.M., Hoyng, S.C., and Kinnard, J.D. "The visual perception of distance ratios outdoors." *Attention, Perception, & Psychophysics* 79, (2017): 1195–1203.

Norman, J.F., Dukes, J.M., Shapiro, H.K., and Peterson, A.E. "The visual perception of large-scale distances outdoors." *Perception* 49, (2020): 968–977.

Norman, J.F., Norman, H.F., Craft, A.E., Walton, C.L., Bartholomew, A.N., Burton, C.L., Wiesemann, E.Y., and Crabtree, C.E. "Stereopsis and aging." *Vision Research* 48, (2008): 2456–2465.

Olson, P.L. and Sivak, M. "Perception-response time to unexpected roadway hazards." *Human Factors* 28, (1986): 91–96.

Ooi, T.L., Wu, B., and He, Z.J. "Distance determined by the angular declination below the horizon." *Nature* 414, (2001): 197–200.

Ortiz-Peregrina, S., Ortiz, C., Casares-López, M., Castro-Torres, J.J., Jimenez del Barco, L., and Anera, R.G. "Impact of age-related vision changes on driving." *International Journal of Environmental Research and Public Health* 17, (2020): 7416–7428.

Oviedo-Trespalacios, O., Truelove, V., Watson, B., and Hinton, J.A. "The impact of road advertising signs on driver behaviour and implications for road safety: A critical systematic review." *Transportation Research Part A* 122, (2019): 85–98.

Owens, D.A., and Tyrrell, R.A. "Effects of luminance, blur, and age on nighttime visual guidance: A test of the selective degradation hypothesis." *Journal of Experimental Psychology: Applied* 5, (1999): 115–128.

Owens, D.A., Wood, J., and Carberry, T. "Effects of reduced contrast on the perception and control of speed when driving." *Perception* 39, (2010): 1199–1215.

Owens, D.A., Wood, J.M., and Owens, J.M. "Effects of age and illumination on night driving: A road test." *Human Factors* 49, (2007): 1115–1131.

Owsley, C., Swain, T., Liu, R., and McGwin, G. "Association of photopic and mesopiccontrast sensitivity in older drivers with risk of motor vehicle collision using naturalistic driving data." *BMC Ophthalmology* 47, (2020). https://doi.org/10.1186/s12886-020-1331-7.

Panerai, F., Droulez, J., Kemeny, A., Balligrand, E., Favre, B., and Berthelot, M. "Speed and safety distance control in truck driving: Comparison of simulation and real-world environment." *Driving Conference DSC 2001* (2001): 91–108. Sophia Antipolis, France.

Peng, J., Wang, C., Shao, Y., and Xu, J. "Visual search efficiency evaluation method for potential connected vehicles on sharp curves." *IEEE Access* 6, (2018): 41827–41838.

Philput, C. "Driver perception of risk: Objective risk versus subjective estimates." In *Proceedings of the Human Factors Society Annual Meeting* 29, (1985): 270–272. Sage, CA: SAGE Publications.

Poulter, D.R. and Wann, J.P. "Errors in motion processing amongst older drivers may increase accident risk." *Accident Analysis & Prevention* 57, (2013): 150–156.

Pradhan, A., Simons-Morton, S., Lee, S., and Klauer, S. "Hazard perception and distraction in novice drivers: Effects of 12 months driving experience." *Proceedings of the Sixth International Driving Symposium on Human Factors in Driver Assessment, Training and Vehicle Design* (June, 2011): 614–620. Lake Tahoe, CA.

Pritchard, S.J. and Hammett, S.T. "The effect of luminance on simulated driving speed." *Vision Research* 52, (2012): 54–60.

Puell, M.C., Palomo, C., Sánchez-Ramos, C., and Villena, C. "Mesopic contrast sensitivity in the presence or absence of glare in a large driver population." *Graefe's Archive for Clinical and Experimental Ophthalmology* 242, (2004): 755–761.

Quimby, A.R. and Watts, G.R. "Human factors and driving performance." No. LR 1004 Monograph, 1981.

Raghuram, A., Lakshminarayanan, V., and Khanna, R. "Psychophysical estimation of speed discrimination. II. Aging effects." *JOSA A* 22, (2005): 2269–2280.

Recarte, M.A. and Nunes L.M. "Perception of speed in an automobile: Estimation and production." Journal of Experimental Psychology: Applied 2, (1996): 291–304.

Ribeiro, F. and Oliveira, J. "Aging effects on joint proprioception: The role of physical activity in proprioception preservation." *European Review of Aging and Physical Activity* 4, (2007): 71–76.

Riby, L., Perfect, T., and Stollery, B. "The effects of age and task domain on dual task performance: A meta-analysis." *European Journal of Cognitive Psychology* 16, (2004): 863–891.

Richardson, E.D. and Marottoli, R.A. "Visual attention and driving behaviors among community-living older persons." *The Journals of Gerontology Series A: Biological Sciences and Medical Sciences* 58, (2003): 832–836.

Rogers, S.D., Kadar, E.E., and Costall, A. "Gaze patterns in the visual control of straight-road driving of speed and experience." *Ecological Psychology* 17, (2005): 19–38.

Romoser, M.R.E., Pollatsek, A., Fisher, D.L., and Williams, C.C. "Comparing the glance patterns of older versus younger experienced drivers: Scanning for hazards while approaching and entering the intersection." *Transportation Research Part F: Traffic Psychology and Behavior* 16, (2013): 104–116.

Rosey, F., Aillerie, I., Espie, S., and Vienna, F. "Driver behaviour in fog is not only a question of degraded visibility: A simulator study." *Safety Science* 95, (2017): 50–61.

Rubin, G.S., Roche, K.B., Prasada-Rao, P., and Fried, L.P. "Visual impairment and disability in older adults." *Optometry and Vision Science* 71, (1994): 750–760.

Rubin, G.S., West, S.K., Munoz, B., Bandeen-Roche, K., Zeger, S., Schein, O., and Fried, L.P. "A comprehensive assessment of visual impairment in a population of older Americans. The SEE Study, Salisbury Eye Evaluation Project." *Investigative Ophthalmology & Visual Science* 38, (1997): 557–68.

Rudin-Brown, C.M. "Vehicle height affects drivers' speed perception: Implications for rollover risk." *Transportation Research Board: Journal of Transportation Research Board* 1899, (2004): 84–89.

Saffarian, M., Happee, R., and de Winter, J.C.F. "Why do drivers maintain short headways in fog? A driving-simulator study evaluating feeling of risk and lateral control during automated and manual car following." *Ergonomics* 55, (2012): 971–985.

Sagberg, F. and Bjornskau, T. "Hazard perception and driving experience and driving experience among novice drivers." *Accident Analysis and Prevention* 38, (2006): 407–414.

Sall, R.J. and Feng, J. "Better off alone: The presence of one hazard impedes detection of another in simulated traffic scenes." *Proceedings of the Human Factors and Ergonomics Society Annual Meeting* (September, 2016): 1444–1448. Washington, DC.

Sall, R.J. and Feng, J. "Dual-target hazard perception: Could identifying one hazard hinder a driver's capacity to find a second?" *Accident Analysis Prevention* 131, (2019): 213–224.

Salthouse, T.A. "The aging of working memory." *Neuropsychology* 8, (1994): 535–543.

Salvucci, D.D. and Gray, R. "A two-point visual control model of steering." *Perception* 33, (2004): 1233–1248.

Sarvi, M. "Heavy commercial vehicles-following behavior and interactions with different vehicle classes." *Journal of Advanced Transportation* 47, (2011): 572–580.

Schieber, F. "Vision and aging." In J.E. Birren and K. Warner Schaie (Eds.) *Handbook of the Psychology of Aging*, 129–161. New York: Academic Press, 2006.

Schiff, W. *Perception: An Applied Approach*. Boston, MA: Houghton Mifflin, 1980.

Schutz, A.C., Billino, J., Bodrogi, P., Polin, D., Khanh, T.Q., and Gegenfurtner, K.R. "Robust underestimation of speed during driving: A field study." *Perception* 44, (2015): 1356–1370.

Scialfa, C.T., Borkenhagen, D., Lyon, J., Deschênes, M., Horswill, M., and Wetton, M. "The effects of driving experience on responses to a static hazard perception test." *Accident Analysis & Prevention* 45, (2012): 547–553.

Scialfa, C.T., Deschênes, M.C., Ference, J.D., Boone, J., Horswill, M.S., and Wetton, M. "Hazard perception in older drivers." *International Journal of Human Factors and Ergonomics* 1, (2012): 221–233

Shaffer, D.M., Maynor, A.B., and Roy, W.L. "The visual perception of lines on the road." *Perception & Psychophysics* 70, (2008): 1571–1580.

Shiffrin, R.M. and Schneider, W. "Automatic and controlled processing revisited." *Psychological Review* 91, (1984): 2697–276.

Shin, H. and Lee, H. "Characteristics of driving reaction time of elderly drivers in the brake pedal task." *Journal of Physical Therapy Science* 24, (2012): 567–570.

Shinar, D. and Scheiber, F. "Visual requirements for safety and mobility of older drivers." *Human Factors* 33, (1991): 507–519.

Shinar, D., McDowell, E.D., and Rockwell, T.H. "Eye movements in curve negotiation." *Human Factors* 19, (1977): 63–71.

Shinar, D., Meir, M., and Ben-Shoham, I. "How automatic is gear shifting?" *Human Factors* 40, (1998): 647–654.

Sidaway, B., Fairweather, M., Sekiya, H., and Mcnitt-Gray, J. "Time-to-collision estimation in a simulated driving task." *Human Factors* 38, (1996): 101–113.

Skaar, N., Rizzo, M., and Stierman, L. "Traffic entry judgments by aging drivers." In *Driving Assessment Conference*, 2. University of Iowa, 2003.

Snowden, R.J. and Kavanagh, E. "Motion perception in the ageing visual system: Minimum motion, motion coherence, and speed discrimination thresholds." *Perception* 35, (2006): 9–24.

Srinivasan, G.N. and Shoba, G. "Statistical texture analysis." *Procedures of World Academy of Science, Engineering and Technology* 26, (2008): 2070–3740.

Staplin, L. "Simulator and field measures of driver age differences in left-turn gap judgments." *Transportation Research Record* 1485, (1995): 49–55.

Stewart, D., Cudworth, C.J., and Lishman, J.R. "Misperception of time-to-collision by drivers in pedestrian accidents." *Perception* (October 1993). https://doi.org/10.1068/p221227.

Stone, L.S. and Thompson, P. "Human speed perception is contrast dependent." *Vision Research* 32, (1992): 1535–1549.

Strasburger, H., Renstschler, I., and Juttner, M. "Peripheral vision and pattern recognition: A review." *Journal of Vision* 11, (2011): 1–82.

Strayer, D.L. and Drews, F.A. "Cell-phone induced inattention blindness." *Current Directions in Psychological Science* 16, (2007): 128–131.

Summala, H. "Accident risk and driver behaviour." *Safety Science* 22, (1996): 103–117.

Summala, H. "Risk control is not risk adjustment: The zero-risk theory of driver behaviour and its implications." *Ergonomics* 31, (1988): 491–506.

Summala, H. "Brake reaction times and driver behavior analysis." *Transportation Human Factors* 2, (2000): 217–226.

Taieb-Maimon, M. and Shinar, D. "Minimum and comfortable driving headways: Reality versus perception." *Human Factors* 43, (2001): 159–172.

Taylor, T.G.G., Masserang, K.M., Pradhan, A.K., Divekar, G., Samuel, S., Muttart, J.W., Pllatsek, A., and Fisher, D.L. "Long term effects of hazard anticipation training on novice drivers measured on the open road." *Proceedings of the International Driving Symposium on Human Factors in Driver Assessment, Training, and Vehicle Design* (2011): 187–194. Iowa City, IA.

Terry, H.R., Charlton, S.G., and Perrone, J.A. "The role of looming and attention capture in drivers' braking responses." *Accident Analysis and Prevention* 40, (July 2008): 1375–1382.

Thorslund, B., Ahlström, C., Peters, B., Eriksson, O., Lidestam, B., and Lyxell, B. "Cognitive workload and visual behavior in elderly drivers with hearing loss." *European Transport Research Review* 6, (2014): 377–385.

Tijerina, L., Kiger, S.M., Rockwell, T.H., and Turnow, C. "Workload assessment of IN-CAB text message system and cellular phone use by heavy vehicle drivers on the road." *Proceedings of the Human Factors and Ergonomics Society Annual Meeting* 39, (1995): 1117–1121.

Tsimhoni, O., Watanabe, H., Green, P., and Friedman, D. "Display of short text messages on automotive HUDs: Effects of driving workload and message location." *University of Michigan Transportation Research Institute.* Technical Report UMTRI-00-13 September, 2000.

Underwood, G., Chapman, P., Brocklehurst, N., Underwood, J., and Crundall, D. "Visual attention while driving: Sequences of eye fixations made by experienced and novice drivers." *Ergonomics* 46, (2003): 629–646.

Underwood, G., Phelps, N., Wright, C., Van Loon, E., and Galpin, A. "Eye fixation scanpaths of younger and older drivers in a hazard perception task." *Ophthalmic and Physiological Optic* 25, (2005): 346–356.

Uno, H. and Hiramatsu, K. "Collision avoidance capabilities of older drivers and improvement by warning presentations." (No. 2001-06-0029). 2001. *SAE Technical Paper.*

Unverricht, J., Samule, S., and Yamani, Y. "Latent hazard anticipation in young drivers: Review and meta-analysis of training studies." Transportation Research Record, Journal of the Transportation Research Board (May, 2018). https:// doi.org/10.1177.

Vaziri-Pashkan, M. and Cavanagh, P. "Apparent speed increases at low luminance." *Journal of Vision* 8, (2008): 1–12.

Venkatraman, V., Lee, J.D., and Schwarz, C.W. "Steer or brake? Modeling drivers'collision avoidance behavior by using perceptual cues." *Transportation Research Record Journal of the Transportation Research Board* (2016). https://doi.org/10.3141/2602-12.

Ventsislavova, P., Gugliotta, A., Pena-Suerez, E., Garcia-Fernandez, P., Eisman, E., Crundall, D., and Castro, C. "What happens when drivers face hazards on the road?" *Accident Analysis and Prevention* 91, (June 2016): 43–54.

Verhaeghen, P. and Cerella, J. "Aging, executive control, and attention: A review of meta-analyses." *Neuroscience & Biobehavioral Reviews* 26, (2002): 849–857.

Verschueren, S.M.P., Brumagne, S., Swinnen, S.P., and Cordo, P.J. "The effect of aging on dynamic position sense at the ankle." *Behavioural Brain Research* 136, (2002): 593–603.

Voelcker-Rehage, C. "Motor-skill learning in older adults – a review of studies on age-related differences." *European Review of Aging and Physical Activity* 5, (2008): 5–16.

Wallace, B. "Driver distraction by advertising: Genuine risk or urban myth?" *Proceedings of the Institute of Civil Engineering ME3*, (2003): 185–190.

Walling, A. and Dickson, G. "Hearing loss in older adults." *American Family Physician* 85, (2012): 1150–1156.

Wallis, T.S.A. and Horswill, M.S. "Using fuzzy signal detection theory to determine why experienced and trained drivers respond faster than novices in a hazard perception test." *Accident Analysis and Prevention* 39, (2007): 1177–1185.

Wan, H., Du, Z., and Yan, Q. "The speed control effect of highway tunnel sidewall markings based on color and temporal frequency." *Journal of Advanced Transportation* 50, (2016): 1352–1365.

Wang, X., Zhu, M., Chen, M., and Tremont, P. "Drivers' rear end collision avoidance behaviors under different levels of situational urgency." *Transportation Research Part C: Emerging Technologies* 71, (2016): 419–433.

Warren, R. "Optical transformation during movement: Review of the optical concomitants of egomotion." 1982-10-01. Ohio State Research Foundation. Columbus, OH, 1982.

Warren, W.H., Arshavir, W., Blackwell, A.W., and Morris, M.W. "Age differences in perceiving the direction of self-motion from optical flow." *Journal of Gerontology* 44, (1989): 147–153.

Werneke, J. and Vollrath, M. "What does the driver look at? The influence of intersection characteristics on attention allocation and driving behavior." *Accident Analysis and Prevention* 45, (2012): 610–619.

Wickens, C.D. "Multiple resources and performance prediction." *Theoretical Issues in Ergonomics Science* 3, (2002): 159–177.

Wickens, C.D. "Multiple resources and mental workload." *Human Factors* 50, (2008): 449–455.

Wilde, G.J.S. "Risk homeostasis theory: An overview." *Injury Prevention* 4, (1998): 89–91.

Wilson, T. and Best, W. "Driving strategies in overtaking." *Accident Analysis and Prevention* 14, (1982): 179–185.

Wittman, M., Kidd, M., Gugg, P., Steffen, A., Fink, M., Poppel, E., and Kamiya, H. "Effects of display position of a visual in-vehicle task on simulated driving." *Applied Ergonomics* 37, (2006): 187–199.

Wood, J.M. "Age and visual impairment decrease driving performance as measured on a closed-road circuit." *Human Factors* 44, (2002): 482–494.

Woods, J.M. and Owens, D.A. "Standard measures of visual acuity do not predict drivers' recognition performance under day or night conditions." *Optometry and Vision Science* 82, (2005): 698–705.

Wu, J., Yang, J., and Yoshitake, M. "Pedal errors among younger and older individuals during different pedal operating conditions." *Human Factors* 56, (2014): 621–630.

Yamani, Y., Horrey, W.J., Liang, Y., and Fisher, D.L. "Age-related differences in vehicle control and eye movement patterns at intersections: Older and middle-aged drivers." *PLoS One* 11(10), (2016): e0164124.

Yan, J.-J., Lorv, B., Li, H., and Sun, H.-J. "Visual processing of the impending collision of a looming object: Time to collision revisited." *Journal of Vision* 11, (2011): 1–25.

Yan, X., Liu, X., Liu,Y., and Jia, Z. "Effects of foggy conditions on drivers speed control behaviors at different risk levels." *Safety Science* 68, (2014): 275–287.

Yan, X., Radwan, E., and Guo, D. "Effects of major road vehicle speed and driver age and gender." *Accident Analysis and Prevention* 39, (2007): 843–852.

Yang, Z., Qiuying, Y., Zhang, W., and Shen, H. "A comparison of experienced and novice drivers' rear-end collision avoidance maneuvers under urgent decelerating events." *Transportation Research Part F: Traffic Psychology and Behaviour* 76, (2012): 353–368.

Yoshimoto, S., Okajima, K., and Takeuchi, T. "Motion perception under mesopic vision." *Journal of Vision* 16, (2016): 1–15.

Young, D., Heckman, G., and Kim, R.. "Human factors in sudden acceleration incidents." In *Proceedings of the Human Factors and Ergonomics Society Annual Meeting* 55, (2011): 1938–1942. Sage, CA: Sage Publications.

Young, M.S., Mahfoud, J.M., Stanton, N.A., Salmon, P.M., Jenkins, D.P., and Walker, G.H. "Conflicts of interest: The implications of roadside advertising for driver attention." *Transportation Research Part F: Traffic Psychology and Behavior* 12, (2009): 381–388.

Yousif, S. and Al-Obaedi, J. "Close following behavior: Testing visual angle car following models using various sets of data." *Transportation Research Part F* (2011): 96–110.

Zheng, Z., Du, Z., Xiang, Q., and Chen, G. "Influence of multiscale visual information on driver's perceived speed in highway tunnels." *Advances in Mechanical Engineering* 10, (2018): 1–12.

Index

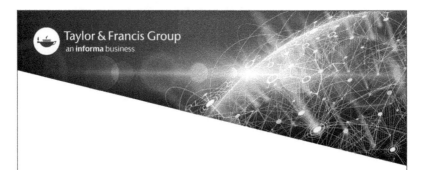

Milton Keynes UK
Ingram Content Group UK Ltd.
UKHW020759121224
451979UK00021B/39

9 781032 431802